Springer Proceedings in Complexity

Springer Proceedings in Complexity publishes proceedings from scholarly meetings on all topics relating to the interdisciplinary studies of complex systems science. Springer welcomes book ideas from authors. The series is indexed in Scopus.

Proposals must include the following:

- name, place and date of the scientific meeting
- a link to the committees (local organization, international advisors etc.)
- scientific description of the meeting
- list of invited/plenary speakers
- an estimate of the planned proceedings book parameters (number of pages/articles, requested number of bulk copies, submission deadline)

Submit your proposals to: Hisako.Niko@springer.com

More information about this series at https://link.springer.com/bookseries/11637

Georgi Yordanov Georgiev
Mahmoud Shokrollahi-Far
Editors

Efficiency in Complex Systems

Self-Organization Towards Increased Efficiency

Springer

Editors
Georgi Yordanov Georgiev
Physics Department
Assumption University
Worcester, MA, USA

Physics Department
Worcester Polytechnic Institute
Worcester, MA, USA

Mahmoud Shokrollahi-Far
University College of Nabi-Akram (UCNA)
Tabriz, Iran

Tilburg University
Tilburg, Noord-Brabant, The Netherlands

ISSN 2213-8684 ISSN 2213-8692 (electronic)
Springer Proceedings in Complexity
ISBN 978-3-030-69287-2 ISBN 978-3-030-69288-9 (eBook)
https://doi.org/10.1007/978-3-030-69288-9

This Springer imprint is published by the registered company Springer Nature Switzerland AG
The registered company address is: Gewerbestrasse 11, 6330 Cham, Switzerland

To my daughter, Victoria; my brother, Iskren; my wife, Boriana; my mother Velichka; and my father, Yordan!

Georgi Yordanov Georgiev

Springer Theses

Recognizing Outstanding Ph.D. Research

Aims and Scope

The series "Springer Theses" brings together a selection of the very best Ph.D. theses from around the world and across the physical sciences. Nominated and endorsed by two recognized specialists, each published volume has been selected for its scientific excellence and the high impact of its contents for the pertinent field of research. For greater accessibility to non-specialists, the published versions include an extended introduction, as well as a foreword by the student's supervisor explaining the special relevance of the work for the field. As a whole, the series will provide a valuable resource both for newcomers to the research fields described, and for other scientists seeking detailed background information on special questions. Finally, it provides an accredited documentation of the valuable contributions made by today's younger generation of scientists.

Theses may be nominated for publication in this series by heads of department at internationally leading universities or institutes and should fulfill all of the following criteria

- They must be written in good English.
- The topic should fall within the confines of Chemistry, Physics, Earth Sciences, Engineering and related interdisciplinary fields such as Materials, Nanoscience, Chemical Engineering, Complex Systems and Biophysics.
- The work reported in the thesis must represent a significant scientific advance.
- If the thesis includes previously published material, permission to reproduce this must be gained from the respective copyright holder (a maximum 30% of the thesis should be a verbatim reproduction from the author's previous publications).
- They must have been examined and passed during the 12 months prior to nomination.
- Each thesis should include a foreword by the supervisor outlining the significance of its content.
- The theses should have a clearly defined structure including an introduction accessible to new PhD students and scientists not expert in the relevant field.

Indexed by zbMATH.

Miranda Louwerse

Efficient Control and Spontaneous Transitions

Doctoral Thesis accepted by Simon Fraser University, Burnaby, BC, Canada

Springer

Miranda Louwerse
Department of Chemistry
Simon Fraser University
Burnaby, BC, Canada

ISSN 2190-5053 ISSN 2190-5061 (electronic)
Springer Theses
ISBN 978-3-031-40533-4 ISBN 978-3-031-40534-1 (eBook)
https://doi.org/10.1007/978-3-031-40534-1

This Springer imprint is published by the registered company Springer Nature Switzerland AG
The registered company address is: Gewerbestrasse 11, 6330 Cham, Switzerland

Paper in this product is recyclable.

Supervisor's Foreword

Spanning three distinct yet related research thrusts, Dr. Miranda Louwerse's PhD thesis addresses deep questions that cut to the physical and informational essence of central chemical quantities such as transition paths and reaction mechanisms, and proposes fundamental new connections between transition-path theory, linear-response theory, nonequilibrium thermodynamics, and information theory. These advances are immediately applicable to design and interpret single-molecule experiment and molecular simulation, and distill engineering principles to guide nano- and biotechnology efforts.

Most prominently, Miranda proved that information theory and nonequilibrium thermodynamics provide robust metrics to identify ideal coarse-grained reaction coordinates that preserve the relevant dynamics of the transition-path ensemble, and that the committor is information-theoretically the perfect reaction coordinate. This gives a thermodynamic measure of the relevance of a degree of freedom to a reaction, providing a practical optimization criterion for identifying reaction coordinates from experiment or simulation of high-dimensional biophysical systems.

Miranda then developed design principles for energetically efficient driving of small stochastic systems. Through extensive numerical simulations that systematically characterized an Ising model, she showed that multi-dimensional control protocols do much less work and keep a driven system significantly closer to equilibrium than their 1D counterparts.

Finally, Miranda directly related such efficient control methods to the spontaneous transition mechanisms in undriven systems. In particular, she discovered that energetically efficient control makes use of pre-existing transition paths, driving degrees of freedom that are activated by spontaneous fluctuations.

Taken as a whole, I consider this PhD thesis to make important advances in several directions, the fruit of impressive creativity and careful characterization. I expect this work will form the conceptual basis for new research directions and point the way to new schemes for computational and experimental inference of dynamic, thermodynamic, and informational aspects of chemical reactions.

Burnaby, BC, Canada
May 2023

Prof. David A. Sivak

Acknowledgements

First, I would like to thank my supervisor David Sivak for accepting a mathematically minded chemist with very little physics background into his research group, and slowly but surely turning me into a theorist. Thank you for being such an involved advisor while simultaneously allowing me to develop independence as a researcher, brainstorming solutions to problems, and sometimes saying just the right thing to spark a new idea whether you realized it or not. Additionally, your focus on clear scientific communication has been a crucial part of my education which I particularly value as I move forward in my career. It has been a true joy to learn from you these past five years.

I thank the members of my supervisory committee over the years, Paul Percival, Michael Eikerling, and Krzysztof Starosta, for their guidance and encouragement that enabled my studies at the intersection of physics and chemistry, and to Paul and Kris for reading and providing valuable feedback on this thesis. I am also grateful for the administrative work by the SFU departments of chemistry and physics that enabled me to straddle the disciplinary line. Additionally, I am immensely grateful for years of financial support from the Natural Sciences and Engineering Research Council, and computational support from the Digital Research Alliance of Canada that enabled the work in this thesis.

The members of the Sivak group over the years have been instrumental in my development as a scientist. They continually amaze me with the mathematical tricks they have up their sleeves and their ability to decompose complex physical phenomena into digestible pieces. Particular thanks to Steve and Jannik for writing the "TAFER paper" which finally made stochastic thermodynamics accessible to me, to Steven for studying minimum-work protocols alongside me for the past few years and listening to my random musings on the topic, and to Emma and Joseph for their friendship and encouragement from the very start. Thanks to Steve, Emma, Joseph, Eric, Matt, Deepak, Jannik, and Matthis for persevering through the pandemic with me and maintaining a strong community despite the lack of proximity. Thank you also to the SFU biophysics community. It has been inspiring to learn from a group of scientists who dabble in all realms of science without regard for traditional disciplinary boundaries. Thank you for your encouragement of early-career researchers and providing multiple avenues for us to get our feet wet and broaden our perspectives on interesting biophysical problems.

Over the years, I have come to a fuller appreciation of the undergraduate education I received at UFV. I particularly thank my undergraduate advisor Noham Weinberg for sparking my interest in chemical physics, for teaching me about statistical mechanics, Brownian motion, and Kramers' theory, and for probing for the first time my (lack of) understanding of reaction coordinates at the microscopic level. I can now confirm that a one-dimensional reaction coordinate does exist and, in theory at least, I know how to find it.

My family and friends have been an endless support. My husband Mitch has been a steady support through the ups and downs of life and research and celebrated victories with me when they do occur. Thank you for taking me outside away from the computer screen to recharge, catch fish, and let my subconscious work while staring at rivers, lakes, and mountains. I am excited to see what the future holds for our life together. Thank you to my parents, siblings, and extended family (both biological and through marriage) for encouraging me throughout this process and keeping me grounded in what truly matters. To my nieces Tessa and Hallie, you are amazing and inspire me every day.

Parts of This Thesis Have Been Published in the Following Journal Articles

M. D. Louwerse and D. A. Sivak, "Information thermodynamics of the transition-path ensemble," *Phys. Rev. Lett.*, **128**, 170602, (2022).

M. D. Louwerse and D. A. Sivak, "Multidimensional minimum-work control of a 2D Ising model," *J. Phys. Chem.*, **156**, 194108, (2022).

M. D. Louwerse and D. A. Sivak, "Connections between efficient control and spontaneous transitions in an Ising model," *Phys. Rev. E*, **106**, 064124, (2022).

Contents

Chapter 1
Introduction

Abstract Quantifying the dynamics and energetics of a system as it undergoes a transition between stable conformations is central to the study of reaction mechanisms and the derivation of reaction rates. For conformational changes in biomolecules, the space the system navigates is high dimensional, presenting challenges to the observation of reactive events in simulation or experiment and obscuring dynamical details that are relevant to the reaction mechanism. Developments in recent decades in transition-path theory, transition-path sampling, and single-molecule experimental techniques have transformed our ability to observe biomolecular reactions in microscopic detail and extract general features of the mechanism. Nearly simultaneously, rapid developments in the field of stochastic thermodynamics have extended familiar notions of work, heat, and entropy to nonequilibrium contexts. One focus of these efforts is the design of protocols that dynamically manipulate the system's conformation with minimal energetic cost. In this thesis, I investigate slow, energetically efficient driving protocols that drive a system between conformations corresponding to endpoints of a reaction, aiming to find connections between principles of efficient driving and the spontaneous transition mechanism in the absence of driving.

The reaction mechanism describing how reactants transform into products is of central interest to chemistry. Complex reactions can be deconstructed into a series of elementary steps whose interdependence leads to the overall reaction kinetics. Additionally, rate-limiting steps that are critical to reaction progress can be identified. Theoretical models describing reaction mechanisms significantly improve understanding of these complex processes, providing language to qualitatively describe the transient behaviour of the system as it undergoes a reaction and quantifying its speed (kinetics) and spontaneous direction (thermodynamics), allowing predictions of how perturbations to the system or environment might change the reaction outcomes.

In biochemical contexts, reactions often consist of conformational changes in large biomolecules [1], where the system must navigate its high-dimensional conformation space between one region characterizing the reactants to another

M. Louwerse, *Efficient Control and Spontaneous Transitions*, Springer Theses,
https://doi.org/10.1007/978-3-031-40534-1_1

characterizing the products. For example, proteins fold from a disordered peptide chain into an energetically stable conformation whose structure and dynamics serve a specific biological function. Molecular machines catalyze unfavorable reactions, pump ions or molecules across cell membranes, and transport cargo around the cell, undergoing significant conformational changes while performing these tasks. These molecules are typically flexible, soft-matter materials in aqueous solution, constantly bombarded with surrounding water molecules that make their motion stochastic. Additionally, biomolecules operate in nonequilibrium environments, consuming chemical energy to fuel a myriad of biological processes and dissipating heat along the way as required by the second law of thermodynamics [1].

Progress in single-molecule experimental techniques, such as atomic force microscopy and optical tweezers [2], has greatly improved the ability to observe biological processes and determine quantitative aspects of their thermodynamics and kinetics [3]. To complement this, computer simulations are routinely performed on large biomolecules, providing atomistic detail of their mechanisms [4, 5].

In this thesis, nonequilibrium driving protocols are investigated [6, 7], which can be used to study conformational changes in biomolecules. If a set of externally controllable parameters are chosen that restrain the system conformation to the reactant region or to the product region, changing these parameters drives the system between conformations of interest. These protocols can be implemented in experiment, for example by holding the two ends of a biomolecule in optical tweezers and pulling them apart or pushing them together to unfold and fold the molecule [8–10], and in simulation, by providing an energetic bias that restrains the system's conformation [11]. Driving protocols provide work to the system and, when performed in finite duration, necessarily dissipate some excess work as heat.

This thesis examines whether there is a correspondence between protocols designed to be energetically efficient when driving the system in finite time between reactant and product, and the reaction mechanism for the spontaneous transition in the absence of driving. Throughout, a classical description of the system that incorporates stochastic dynamics is used, which is well-suited to describing conformational changes in biomolecules in noisy environments [1, 2]. Additionally, a reaction is described generally as a transition from a reactant molecule or conformation A to product B in order to represent the concept of a reaction without details of the system involved.

1.1 Spontaneous Transitions

How do reactions occur? The reactant and product are usually metastable molecular species or conformations, meaning their free energy is low relative to the free-energetic barrier separating them (Fig. 1.1). In order for the reaction to occur, the system must accumulate sufficient free energy to overcome the barrier; reaction theories attempt to quantify the frequency of this event by considering the underlying dynamics and energetics.

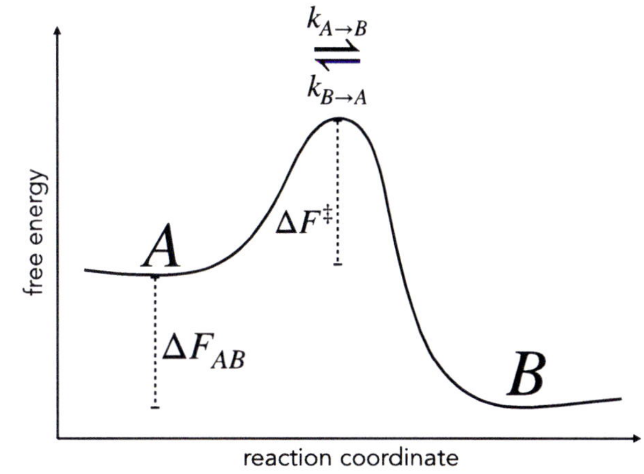

Fig. 1.1 Schematic for an $A \to B$ reaction. The system free energy is plotted as a function of the reaction coordinate, with metastable states A and B separated by a free-energy barrier. The reaction free energy is the difference between A and B, ΔF_{AB}. The frequency with which the system crosses the energy barrier from $A \to B$ ($B \to A$) is given by rate constants $k_{A\to B}$ and $k_{B\to A}$

The speed and spontaneous direction of a reaction from reactant A to product B are parameterized by phenomenological rate constants $k_{A\to B}$ and $k_{B\to A}$ and equilibrium free-energy difference ΔF_{AB} [12]

$$A \underset{k_{B\to A}}{\overset{k_{A\to B}}{\rightleftharpoons}} B\,. \tag{1.1}$$

Finding values for these reaction parameters either experimentally, theoretically, or computationally, is a major focus of research in quantitative sciences. Importantly, the rate constants and free-energy difference are related by the principle of detailed balance [13]

$$\frac{k_{A\to B}}{k_{B\to A}} = e^{-\beta\Delta F_{AB}}\,, \tag{1.2}$$

providing a constraint on the equilibrium kinetic and thermodynamic parameters of the process.

1.1.1 Reaction Theories

One of the first conceptual insights into chemical reactions is the existence of an energetic barrier separating metastable states A and B; a successful reaction requires the system to gain sufficient *activation energy* to form the *activated complex* or *transition-state species* at the peak of the energy barrier, then release this energy as it forms the stable product. The activation energy relates the rate of a reaction to the temperature of the environment through the Arrhenius equation [12]

$$k_{A\to B} = \Gamma e^{-\beta\Delta E^{\ddagger}}\,, \tag{1.3}$$

for inverse temperature $\beta = 1/k_B T$ in units of Boltzmann's constant, and energy difference $\Delta E^{\ddagger} \equiv E_{TS} - E_A$ between the activated complex (transition state) and the reactant. This explains a qualitative observation about reactions: as the temperature is increased, the speed of the reaction increases. Measurements of the rate constant at a range of temperatures allow calculation of the activation energy. The prefactor Γ is referred to as a frequency factor, accounting for the frequency of collisions that allow reactant A to form the activated complex. For reactions in the gas phase, this frequency factor can be accounted for via collision theory [12] which estimates the frequency of collisions with sufficient energy and correct orientation for the system to overcome the energy barrier at a particular temperature. For conformational changes in solution, collisions come from the surrounding solvent molecules and the prefactor can be estimated using transition-state theory and a diffusion equation for the dynamics [14].

Transition-state theory derives the rate constant using statistical mechanics [15–17], providing a connection between the rate constant and thermodynamic parameters of the activated complex. The transition state is imagined as a hypersurface of the configuration space parameterized by a reaction coordinate that the system must pass through as it reacts from A to B. Assuming that the reactant region is in thermal equilibrium and forbidding any recrossing of the transition-state surface, statistical mechanics can be used to derive the rate constant in terms of the *free-energy* difference between the reactant A and the transition state,

$$k_{A\to B}^{\mathrm{TST}} = \frac{k_B T}{h} e^{-\beta \Delta F^{\ddagger}} = \frac{k_B T}{h} e^{\Delta H^{\ddagger}/k_B} e^{-\beta \Delta E^{\ddagger}} . \tag{1.4}$$

where $\Delta H^{\ddagger}$ is the activation entropy of the reaction, accounting for the difference in the entropy of the accessible conformational space in the transition-state region and the reactant region, and h is Planck's constant. Transition-state theory has been wildly successful at describing a wide variety of reactions and extended to include quantum effects such as tunnelling [18]. Its language and concepts form the basis of a chemist's understanding of activated processes.

A drawback of transition-state theory is the absence of recrossing; once the system has passed through the transition-state surface from A it is assumed to never return. Thus the unidirectional flux across the barrier used to derive the rate constant in transition-state theory overestimates the true rate constant,

$$k_{A\to B} = \kappa k_{A\to B}^{\mathrm{TST}} \tag{1.5}$$

where $\kappa \leq 1$ is a transmission coefficient correcting for recrossing events [19]. The transmission coefficient and the activation free energy each depend on the choice of dividing surface (although their product is always the "true" rate constant). By varying the position of the dividing surface away from the top of the energy barrier, different estimates of the rate constant are obtained via transition-state theory (which excludes calculation of the transmission coefficient). For example, variational transition-state theory [18, 19] finds a position for the transition-state surface that minimizes the rate constant estimated using transition-state theory.

However, consideration of recrossing events is necessary to estimate the transmission coefficient and the "true" rate constant. Ideally, its calculation should consider both the immediate dynamics at the transition state and information about the trajectory at later times so that trajectories that cross the transition-state surface multiple times are counted a single time toward the reactive flux. An improvement on the transition-state theory assumptions was made by Kramers [14, 20] to allow calculation of the transmission coefficient for dynamics described by diffusion on a one-dimensional reaction coordinate, considering both high- and low-friction regimes. Assuming that the system is in local equilibrium in the energetic well corresponding to mesostate A (meaning the initial probability distribution in the state space on A's side of the transition barrier is proportional to the Boltzmann distribution), Kramers assumes a steady state where once the system reaches the product it is reinitialized in the reactant state, generating a flux of trajectories from A to B. The rate at which system probability leaks over the barrier into mesostate B can be estimated from the diffusion equation by approximating the free-energy surface near A and the transition-state region as quadratic functions with curvatures ω_A and $\omega^{\ddagger}$ respectively. In the high-friction limit, this yields the transmission coefficient

$$\kappa_{\text{Kramers}} = \frac{\omega_A \omega^{\ddagger}}{2\pi\gamma} \tag{1.6}$$

as a correction to the transition-state theory rate constant that accounts for recrossing events, where γ is the friction coefficient for the diffusive system.

Since Kramers' reaction theory models the noisy environment encountered by bio-molecules, it is useful in predicting rate constants for biomolecular transformations so long as the dynamics are accurately described by diffusion on a one-dimensional reaction coordinate. For example, it has been useful in interpreting information from atomic force spectroscopy experiments [21, 22], where, e.g., protein folding can be probed by pulling the ends of the molecule apart. However, the *a priori* assumption of a one-dimensional reaction coordinate that accurately captures the reaction dynamics is challenging since criteria for identifying such a coordinate are unclear [23]. Kramers' description of diffusion-dominated activated processes has been extended to multidimensional diffusion [24], easing the assumption of a one-dimensional reaction coordinate, but still requiring knowledge of a small set of relevant coordinates to make the calculation tractable. However, the development of transition-path theory provides a quantitative definition of the "true" one-dimensional reaction coordinate required for Kramers' description, as discussed in Sect. 1.1.2.

1.1.2 Transition-Path Theory

While the concept of a transition state is useful in many contexts, there are also many reactions that cannot be described using transition-state theory. These include conformational changes where the energy landscape is rugged or lacks an identifiable transition state that is a bottleneck for the reaction. Additionally, biomolecules have very high-dimensional conformational spaces (in principle, $3N$ coordinates are required to describe the three-dimensional position of an N-atom molecule), resulting in massive amounts of data that require dimensionality reduction to extract useful information, such as a small set of coordinates that capture important aspects of the reaction mechanism. It is usually not clear *a priori* what system coordinates might serve as good reaction coordinates, despite the fact that reaction theories tend to rely on their existence and appropriate identification. It is therefore important to derive a reaction theory that does not depend on already knowing a physical one-dimensional coordinate or any specific details about the reaction mechanism, including the presence of a single transition state. This allows the theory to describe many types of reactions, either as an activated process with a single energy barrier or as diffusion through a network of mesostates as in Markov-state models [25]. Seminal work in the last few decades in transition-path sampling [26–28] and transition-path theory [29, 30] have come far in quantitatively describing reactions using statistical tools, also yielding a one-dimensional reaction coordinate between any two mesostates A and B.

It is first important to understand some of the challenges associated with describing reactions in atomistic detail, rather than integrating out degrees of freedom deemed irrelevant to the reaction. Of particular relevance to this discussion is classical molecular dynamics simulations, which describe a molecule's three-dimensional position and velocity as the system is propagated forward in time using Newton's equations of motion, with the system's potential energy described by an empirical force field which sums energetic contributions from internal degrees of freedom (bonds, angles, and torsions) as well as Coulombic and interatomic repulsive forces between pairs of atoms. Molecular dynamics simulations are widely used throughout the natural sciences to study a variety of problems in biology, chemistry, and materials, elucidating relationships between the structure, function, and dynamics of molecules.

To ensure accuracy in simulating the dynamics of many interacting atoms, a relatively small integration time step must be used, typically on the order of femtoseconds; in contrast, timescales for biologically relevant reactions are typically on the order of micro-seconds to seconds [25]. This timescale separation means that gathering samples of the rare *reactive trajectories* of the system that travel between metastable states in order to determine their statistics cannot be done with "brute force" by simply simulating the dynamics until reactions occur; instead, techniques must be developed to directly sample these trajectories. The same timescale problem will appear again when discussing collective variables and free-energy calculations in Sect. 2.3.1.

The seminal method for directly sampling reactive trajectories was developed in 1998 by Dellago et al. [26], known as *transition-path sampling*. Since then, many more methods to sample and analyze transition paths have been developed, such as transition-interface sampling [31], milestoning [28], and forward-flux sampling [32]. The original transition-path sampling technique enacts a Markov process in the space of trajectories, using random shooting from a trial trajectory to generate new samples. From this, a sample of trajectories from A to B is obtained, which can be used to obtain rate constants using Chandler's reactive-flux formalism [26, 33]. In principle, information about the reaction mechanism is also contained in this trajectory sample, but a theoretical framework is needed to extract and interpret such information.

Thus transition-path theory [29] was developed to quantitatively describe the entire reaction and determine its rate constant, without assumptions of metastability for the reactant and product or any specific details of the reaction mechanism (e.g., the presence of a single transition state). Imagine a long ergodic equilibrium trajectory that travels infinitely many times between two non-overlapping subsets of microstates corresponding to the reactant and product, A and B. Next, select from this ergodic supertrajectory all the trajectory segments that leave A and next reach B without returning to A. Each trajectory segment is a *transition path* or *reactive trajectory*, and the entire ensemble of reactive trajectories is the *transition-path ensemble* (TPE). All reaction parameters as well as the reaction mechanism can be determined from this ensemble of trajectories.

The statistical description of the transition-path ensemble relies on the definition of the *committor function* $q(\boldsymbol{\phi})$ (also called the commitment or splitting probability [34], or the folding probability in the protein folding literature [35]), the probability that a trajectory initiated from a microstate $\boldsymbol{\phi}$ next reaches B before A. The committor maps the conformation space onto the interval $q(\boldsymbol{\phi}) \in [0, 1]$, identifying the transition-state ensemble as conformations making up the $q(\boldsymbol{\phi}) = 0.5$ isocommittor surface [36]. The committor allows calculation of the reaction rate, mean first-passage time, and mean transition-path duration from a one-dimensional description [37] providing a connection to Kramers' theory and the required one-dimensional reaction coordinate. The versatility of transition-path theory in describing a wide variety of systems makes it a leading theory today in determining reaction mechanisms in computational sciences. Additionally, transition-path theory has been used to interpret experimental data and estimate the committor function for a protein-folding reaction [3, 38]. Some quantitative and qualitative aspects of this theory are reviewed further in Sect. 2.2.

The committor's relevance to determining reaction parameters and reaction mechanisms has earned it the title of the "true" or "ideal" one-dimensional reaction coordinate [23, 39–42]. Its importance to chemical physics has spurred work to use machine learning to estimate the committor for systems of interest [43–45] (calculating the committor is generally a computationally expensive task [23]). However, the committor does not necessarily convey mechanistically useful information, as it simply gives a number for each microstate of the system. Instead, the dominant features of the system on the isocommittor surfaces (all conformations with the same committor value) [46] are of interest. In particular, it is helpful to determine a small set of physically relevant collective variables (functions of system microstate) that provide intuitive understanding about the reaction mechanism; the committor and/or other statistics (such as the free-energy landscape) can then be determined in the given collective-variable space. To obtain accurate information about the reaction mechanism, the chosen set of collective variables must capture the relevant information about the reaction. A number of approaches have been proposed to identify small sets of relevant collective variables, often relying on ideas of timescale separation to identify slow degrees of freedom by approximating eigenfunctions of the dynamical operator [47, 48] or determining a reduced representation that is predictive about future states of the system [49]. The development of methods to identify and represent relevant degrees of freedom for a reaction is currently an active area of research.

1.2 Efficient Control

Considerable progress has been made in recent decades in understanding nonequilibrium systems and the statistical nature of the second law of thermodynamics, including the derivation of nonequilibrium work relations, fluctuation theorems, and thermodynamic uncertainty relations [6, 50–53]. In this thesis, control protocols which drive the system between conformations corresponding to the endpoints A and B of a reaction are investigated. In contrast to transition paths, which occur due to spontaneous fluctuations from the environment occasionally giving rise to rare reactive trajectories, control protocols are enacted by an experimenter external to the system, who changes a set of *control parameters* according to a fixed schedule, driving the system between thermodynamic states which could be designed to correspond to reaction endpoints. Control protocols can be implemented in experiment [8–10, 54] or simulation [55, 56], providing molecular-level information about a wide variety of systems.

1.2.1 Driving Protocols in Simulation and Experiment

Control parameters can be any externally tunable parameters of the system that allow the system to be driven between desired thermodynamic states. Examples from classical thermodynamics include temperature, pressure, and volume. Single-molecule experimental techniques developed over the last decades have broadened the spectrum of available control parameters. For example, optical tweezers can provide force to the ends of a protein and cause it to unfold; in this case the distance between optical tweezers holding the ends of the molecule serves as a control parameter, and it is tuned according to some protocol to drive the system between folded and unfolded conformations. Significant experimental and theoretical work has been done to learn thermodynamic properties of systems from their response to such experimental control protocols [22, 57, 58]. Computer simulations offer even greater flexibility in choosing control parameters; essentially any function of the system's atomic coordinates can be driven between relevant endpoints via some time-dependent biasing potential.

This thesis investigates protocols that minimize the average excess work done on the system during a long-but-finite duration, and probes their relationship to spontaneous transition paths. In order to explain some relevant concepts related to control protocols, imagine an ideal gas contained in a piston with adjustable volume that is driven from a low-pressure to high-pressure state by decreasing the volume in a constant-temperature environment (volume is the control parameter in this scenario).

If the volume is changed slowly throughout the protocol (a reversible or quasistatic process), the system relaxes continuously to remain in thermal equilibrium with the environment, and the distribution of gas molecules throughout the piston remains uniform. Work is done on the system reversibly, equal to the change in free energy between the final and initial equilibrium ensembles corresponding to low- and high-pressure states of the system [12]

$$\mathcal{W}^{\mathrm{rev}} = \Delta F^{\mathrm{eq}} = k_{\mathrm{B}} T \ln \frac{V_{\mathrm{f}}}{V_{\mathrm{i}}} , \tag{1.7}$$

where V_{i} and V_{f} are the initial and final volumes of the system, respectively.

However, if the volume is reduced relatively quickly, the gas particles will not instantaneously be uniformly distributed throughout the piston; they only partially relax towards the uniform distribution throughout the control protocol, and only fully equilibrate when the system is left for a long time at the final volume. The system is out of equilibrium with the environment throughout the protocol, and on average excess work is performed on the system above the equilibrium free-energy change:

$$\langle \mathcal{W}^{\mathrm{ex}} \rangle_{\Lambda} \equiv \langle \mathcal{W} \rangle_{\Lambda} - \Delta F^{\mathrm{eq}} \geq 0 , \tag{1.8}$$

where $\langle \ldots \rangle_{\Lambda}$ denotes an average over system responses to the same driving protocol Λ. The excess work is dissipated as heat to the environment when the system relaxes, quantifying the energy "wasted" during the process. This inequality for excess work derives from the second law of thermodynamics, which requires irreversible entropy production for any nonequilibrium process.

The second law of thermodynamics is a statistical law that is only true on average [51]. For systems such as biomolecules that experience large fluctuations, the excess work done on the system differs each time the protocol is performed, forming an excess-work distribution $P(\mathcal{W}^{\mathrm{ex}}|\Lambda)$. This distribution can have nonzero probability for negative values of excess work, indicating that for some trajectories the work is less than the free-energy difference. However, the average excess work is positive (1.8).

Remarkably, the inequality Eq. (1.8) can be derived from an equality (assuming the system begins in equilibrium),

$$\langle e^{-\beta \mathcal{W}} \rangle_{\Lambda} = e^{-\beta \Delta F^{\text{eq}}} , \tag{1.9}$$

which was first introduced by Jarzynski in 1997 [50]. The Jarzynski equation holds for any protocol duration (even when the system is far from equilibrium during the protocol), in principle allowing calculation of the equilibrium free-energy difference between thermodynamic states from nonequilibrium system responses. Soon after, Crooks derived a work fluctuation theorem for nonequilibrium control protocols [6]

$$\frac{P(\mathcal{W}|\Lambda)}{P(-\mathcal{W}|\tilde{\Lambda})} = e^{-\beta \mathcal{W}^{\text{ex}}} , \tag{1.10}$$

which relates the work distributions in the forward and time-reversed protocols to the excess protocol work. The time-reversed protocol begins with the system in equilibrium at the final control parameter, then runs the protocol in reverse to reach the initial control parameter (e.g., expanding the gas in the piston rather than compressing it). When the process is performed in finite time, the system lags the equilibrium distribution during both the forward and reverse protocols, yielding different work distributions for each process. However, Eq. (1.10) places a tight constraint on the allowable work distributions for the forward and time-reversed processes in terms of the excess work. Similar fluctuations theorems have been derived for nonequilibrium processes other than control protocols [51], relating the probabilities of forward and time-reversed processes to irreversible entropy production.

These nonequilibrium work relations allow equilibrium thermodynamic information to be extracted from nonequilibrium driving protocols. For example, the free-energy difference between equilibrium ensembles can be computed using Eq. (1.9), although in practice bidirectional variations on (1.9) such as Bennett's acceptance ratio [59] are used to improve accuracy and reduce bias in the estimate [60–62]. The potential of mean force on the system along the driving coordinate can also be calculated using nonequilibrium protocols, as shown by Hummer and Szabo [63]. This makes nonequilibrium protocols a useful tool for determining quantitative information about the system's underlying free-energy landscape. The potential of mean force for DNA hairpin folding has been calculated experimentally using such methods [10].

1.2.2 *Minimum-Work Protocols*

The equilibrium free-energy difference between control-protocol endpoints is fixed because free energy is a state function, however the work done during the protocol is path dependent. For a fixed protocol duration, it is therefore of interest to find a protocol that minimizes excess work. Such minimum-work protocols (MWPs) in turn provide insight into energetically efficient operation of biological molecular machines [1], the design of methods for efficient chemical processing [64], and informs the design of novel nanotechnology [65]. Additionally, the magnitude of excess work has implications for the precision and accuracy of free-energy estimates from nonequilibrium control protocols [60–62], which are improved when less excess work is done. Exact results for minimum-work protocols for any duration have been derived in a few model systems [66]; approximations are available in the limits of long and short protocol duration [7, 67–70].

In this thesis, minimum-work protocols in the limit of long-but-finite duration are considered, as derived in Ref. [7]. Here, a linear-response expansion of the system's response to the driving protocol yields a Riemannian metric in control-parameter space measuring the excess power during the protocol. Minimum-work protocols designed using this metric are geodesics, the generalization of straight lines between protocol endpoints to curved, non-Euclidean geometry.

The metric is interpreted as generalized friction, quantifying system resistance (due to equilibrium fluctuations) to changes in the control parameter. For the example of isothermal compression of gas in a piston, this resistance comes from the particles accumulating near the piston as the volume is reduced because there has not been sufficient time for the particles to diffuse out of the way and form the equilibrium uniform distribution. This same idea is intuitive in biomolecular experiments, for example in optical-tweezer experiments where a protein is pulled apart. When the pulling is done in finite time, interactions within the protein that keep it in a folded state resist the changes in control parameter that drive these interactions to break.

The generalized friction encodes information about changes in the system's equilibrium free-energy during the protocol [67] (its thermodynamics) and equilibrium relaxation modes [68, 71] (its kinetics). Since generalized friction is related to equilibrium fluctuations, it seems natural that the geometry of control-parameter space would reflect the geometry of the underlying system energetics and dynamics.

1.3 Motivation for This Thesis

To study the correspondence between driving protocols and spontaneous transitions, control protocol endpoints are chosen such that the system is restrained to reaction endpoints A or B when in equilibrium at corresponding control-parameter values. The choice of A and B and appropriate control parameters depend on the system of interest; in this thesis, calculations are performed on a particular model system but general principles will be drawn that could apply to many systems of interest.

The long-duration limit for driving protocols is considered rather than studying minimum-work protocols for all durations. The generalized friction metric that derives from a linear-response approximation of a protocol's excess work has enticing features that may enable a correspondence with spontaneous transition mechanisms. It quantifies resistance to driving the system through an ensemble of pathways in conformational space by capturing local features of the system's free energy and relaxation times in the corresponding conjugate-force space, which reflects the dynamics of collective variables the control parameters are able to control. Local features of the system's energetics and dynamics are also relevant to characterizing transient dynamics during a spontaneous transition path [72–75]. If there is a significant energy barrier the system must overcome to get from A to B, it seems intuitive that a minimum-work protocol would drive the system through a saddle point on the free-energy surface (where the transition state of the spontaneous reaction is expected to occur), requiring the minimum work possible to drive the system over the energy barrier. This suggests a correspondence between the minimum-work protocol and the spontaneous transition path of the system between A and B which will be investigated throughout this thesis.

Theoretical descriptions of minimum-work protocols and spontaneous transitions differ in several ways. Minimum-work protocols are performed in a fixed duration, while spontaneous transition paths from A to B have a range of duration. Protocols exchange work with the system throughout the protocol, while there is no work performed on the system during a spontaneous transition. Protocol endpoints are defined in control-parameter space, while endpoints of a spontaneous transition are defined in conformational (or collective-variable) space. Protocols drive the system out of equilibrium, while spontaneous transitions occur when the system is at equilibrium.

Nevertheless, there are some commonalities between the processes that are worth investigating. Both processes share the same conformational space, with internal energy coupling the system's many degrees of freedom. The energy and entropy of the conformational space, and changes in these state functions during trajectories either generated by a control protocol or selected as a transition path, can therefore be directly compared. With appropriate choice of control parameters, the endpoint distributions of the protocol can approximate the unperturbed distribution in metastable basins surrounding transition-path endpoints A and B, providing at minimum a control-parameter space capable of distinguishing metastable conformations. However, there is great interest in sets of collective variables that not only distinguish metastable basins but also provide information about the reaction mechanism. A control-parameter space is designed to bias collective variables that are presumed to be relevant to a reaction and minimum-work protocols are found and tested for their correspondence to the spontaneous reaction mechanism.

1.4 Outline of This Thesis

Chapter 2 begins with some mathematical background on stochastic dynamics, transition-path theory, and minimum-work protocols. Chapter 3 provides a thermodynamic accounting of the information contained in reactive trajectories, allowing quantitative determination of the relevance of a set of collective variables to a reaction. This establishes a connection between a dynamical theory describing spontaneous transitions and thermodynamic definitions of energy and entropy along individual trajectories [76]. In Chap. 4, minimum-work protocols for driving an Ising model system between energetically stable configurations are calculated. The additional freedom provided by controlling multiple collective variables is explored and related to reduced excess work relative to a naive (constant-velocity) protocol between the same endpoints [77]. In Chap. 5, the two paradigms are drawn together, comparing the mechanism and energy flows into the system during a minimum-work protocol and during the spontaneous transition-path ensemble, highlighting some intuitive connections between the processes and some complicating factors. Conclusions and suggestions for future work are discussed in Chap. 6.

References

1. Brown, A.I., Sivak, D.A.: Theory of nonequilibrium free energy transduction by molecular machines. Chem. Rev. **120**(1), 434–459 (2020)
2. Yanagida, T.: Fluctuation as a tool of biological molecular machines. BioSystems **93**(1–2), 3–7 (2008)
3. Neupane, K., Hoffer, N.Q., Woodside, M.T.: Measuring the local velocity along transition paths during the folding of single biological molecules. Phys. Rev. Lett. **121**, 018102 (2018)
4. Karplus, M., McCammon, J.A.: Molecular dynamics simulations of biomolecules. Nat. Struct. Biol. **9**(9), 646–652 (2002)
5. Hartmann, C., Banisch, R., Sarich, M., Badowski, T., Scáž§utte, C.: Characterization of rare events in molecular dynamics. Entropy **16**(1), 350–376 (2014)
6. Crooks, G.E.: Path-ensemble averages in systems driven far from equilibrium. Phys. Rev. E **61**(3), 2361–2366 (2000)
7. Sivak, D.A., Crooks, G.E.: Thermodynamic metrics and optimal paths. Phys. Rev. Lett. **108**(19), 190602 (2012)
8. Liphardt, J., Dumont, S., Smith, S.B., Tinoco, I., Bustamante, C.: Equilibrium information from nonequilibrium measurements in an experimental test of Jarzynski's equality. Science **296**, 1832–1835 (2002)
9. Tafoya, S., Large, S.J., Liu, S., Bustamante, C., Sivak, D.A.: Using a system's equilibrium behavior to reduce its energy dissipation in nonequilibrium processes. Proc. Natl. Acad. Sci. U. S. A. **116**(13), 5920–5924 (2019)
10. Engel, M.C., Ritchie, D.B., Foster, D.A., Beach, K.S., Woodside, M.T.: Reconstructing folding energy landscape profiles from nonequilibrium pulling curves with an inverse Weierstrass integral transform. Phys. Rev. Lett. **113**(23), 1–5 (2014)
11. Moradi, M., Tajkhorshid, E.: Computational recipe for efficient description of large-scale conformational changes in biomolecular systems. J. Chem. Theory Comput. **10**(7), 2866–2880 (2014)
12. Atkins, P., de Paula, J., Keeler, J.: Physical Chemistry, 11th edn. Oxford University Press, Oxford (2018)
13. Zwanzig, R.: Nonequilibrium Statistical Mechanics. Oxford University Press, New York (2001)
14. Kramers, H.A.: Brownian motion in a field of force and the diffusion model of chemical reactions. Physica **7**(4), 284–304 (1940)
15. Eyring, H.: The activated complex in chemical reactions. J. Chem. Phys. **3**, 107–115 (1935)
16. Evans, M.G., Polanyi, M.: Some applications of the transition state method to the calculation of reaction velocities, especially in solution. Trans. Faraday Soc. **31**, 875 (1935)

17. Wigner, E.: Calculation of the rate of elementary association reactions. J. Chem. Phys. **5**(0), 720–725 (1937)
18. Bao, J.L., Truhlar, D.G.: Variational transition state theory: theoretical framework and recent developments. Chem. Soc. Rev. **46**(24), 7548–7596 (2017)
19. Peters, B.: Chapter 10 - Transition state theory. In: Peters, B. (ed.) Reaction Rate Theory and Rare Events Simulations, pp. 227–271. Elsevier, Amsterdam (2017)
20. Peters, B.: Chapter 16 - Kramers theory. In: Peters, B. (ed.) Reaction Rate Theory and Rare Events Simulations, pp. 435–450. Elsevier, Amsterdam (2017)
21. Dudko, O.K.: Decoding the mechanical fingerprints of biomolecules. Q. Rev. Biophys. **49**(e3), 1–14 (2016)
22. Dudko, O.K., Hummer, G., Szabo, A.: Theory, analysis, and interpretation of single-molecule force spectroscopy experiments. PNAS **105**(41), 15755–15760 (2008)
23. Peters, B.: Reaction coordinates and mechanistic hypothesis tests. Annu. Rev. Phys. Chem. **67**, 669–690 (2016)
24. Langer, J.: Statistical theory of the decay of metastable states. Ann. Phys. (N. Y). **54**(2), 258–275 (1969)
25. Chodera, J.D., Noé, F.: Markov state models of biomolecular conformational dynamics. Curr. Opin. Struct. Biol. **25**, 135–144 (2014)
26. Dellago, C., Bolhuis, P.G., Csajka, F.S., Chandler, D.: Transition path sampling and the calculation of rate constants. J. Chem. Phys. **108**, 1964 (1998)
27. Bolhuis, P.G., Chandler, D., Dellago, C., Geissler, P.L.: Transition path sampling: throwing ropes over rough mountain passes, in the dark. Annu. Rev. Phys. Chem. **53**, 291–318 (2002)
28. Faradjian, A.K., Elber, R.: Computing time scales from reaction coordinates by milestoning. J. Chem. Phys. **120**, 10880–10889 (2004)
29. Weinan, E., Vanden-Eijnden, E.: Towards a theory of transition paths. J. Stat. Phys. **123**(3), 503–523 (2006)
30. Metzner, P., Schutte, C., Vanden-Eijnden, E.: Transition path theory for Markov jump processes. Multiscale Model. Simul. **7**(3), 1192–1219 (2009)
31. Van Erp, T.S., Moroni, D., Bolhuis, P.G.: A novel path sampling method for the calculation of rate constants. J. Chem. Phys. **118**, 6617 (2003)
32. Allen, R.J., Warren, P.B., Ten Wolde, P.R.: Sampling rare switching events in biochemical networks. Phys. Rev. Lett. **94**, 018104 (2005)
33. Peters, B.: Chapter 13 - Reactive flux. In: Peters, B. (ed.) Reaction Rate Theory and Rare Events Simulations, pp. 335–362. Elsevier, Amsterdam (2017)
34. Onsager, L.: Initial recombination of ions. Phys. Rev. **54**(8), 554–557 (1938)
35. Du, R., Pande, V.S., Grosberg, A.Y., Tanaka, T., Shakhnovich, E.: On the transition coordinate for protein folding. J. Chem. Phys. **108**, 334–350 (1998)
36. Berezhkovskii, A., Szabo, A.: One-dimensional reaction coordinates for diffusive activated rate processes in many dimensions. J. Chem. Phys. **122**, 014503 (2005)
37. Berezhkovskii, A.M., Szabo, A.: Diffusion along the splitting/commitment probability reaction coordinate. J. Phys. Chem. B **117**, 13115–13119 (2013)
38. Pyo, A.G.T., Hoffer, N.Q., Neupane, K., Woodside, M.T.: Transition-path properties for folding reactions in the limit of small barriers. J. Chem. Phys. **149**, 115101 (2018)
39. Weinan, E., Vanden-Eijnden, E.: Transition-path theory and path-finding algorithms for the study of rare events. Annu. Rev. Phys. Chem. **61**, 391–420 (2010)
40. Li, W., Ma, A.: Recent developments in methods for identifying reaction coordinates. Mol. Simul. **40**(10–11), 784–793 (2014)
41. Peters, B., Bolhuis, P.G., Mullen, R.G., Shea, J.-E.: Reaction coordinates, one-dimensional Smoluchowski equations, and a test for dynamical self-consistency. J. Chem. Phys. **138**, 054106 (2013)
42. Banushkina, P.V., Krivov, S.V.: Optimal reaction coordinates. WIREs Comput. Mol. Sci. **6**, 748–763 (2016)
43. Khoo, Y., Lu, J., Ying, L.: Solving for high-dimensional committor functions using artificial neural networks. Res. Math. Sci. **6**, 1 (2019)

44. Li, Q., Lin, B., Ren, W.: Computing committor functions for the study of rare events using deep learning. J. Chem. Phys. **151**, 54112 (2019)
45. Rotskoff, G.M., Mitchell, A.R., Vanden-Eijnden, E.: Active importance sampling for variational objectives dominated by rare events: consequences for optimization and generalization. In: Mathematical and Scientific Machine Learning PMLR, pp. 757–780 (2022)
46. Bolhuis, P.G., Dellago, C.: Practical and conceptual path sampling issues. Eur. Phys. J. Spec. Top. **224**, 2409–2427 (2015)
47. Pérez-Hernández, G., Paul, F., Giorgino, T., De Fabritiis, G., Noé, F.: Identification of slow molecular order parameters for Markov model construction. J. Chem. Phys. **139**(1), 015102 (2013)
48. Hernández, C.X., Wayment-Steele, H.K., Sultan, M.M., Husic, B.E., Pande, V.S.: Variational encoding of complex dynamics. Phys. Rev. E **97**(6), 1–11 (2018)
49. Wang, Y., Tiwary, P.: State predictive information bottleneck. J. Chem. Phys. **154**, 134111 (2021)
50. Jarzynski, C.: Nonequilibrium equality for free energy differences. Phys. Rev. Lett. **78**(14), 2690–2693 (1997)
51. Seifert, U.: Stochastic thermodynamics, fluctuation theorems and molecular ma- chines. Rep. Prog. Phys. **75**, 126001 (2012)
52. Barato, A.C., Seifert, U.: Thermodynamic uncertainty relation for biomolecular processes. Phys. Rev. Lett. **114**(15), 1–5 (2015)
53. Gingrich, T.R., Horowitz, J.M., Perunov, N., England, J.L.: Dissipation bounds all steady-state current fluctuations. Phys. Rev. Lett. **116**(12), 1–5 (2016)
54. Gupta, A.N., Vincent, A., Neupane, K., Yu, H., Wang, F., Woodside, M.T.: Experimental validation of free-energy-landscape reconstruction from non-equilibrium single-molecule force spectroscopy measurements. Nat. Phys. **7**(8), 631–634 (2011)
55. Dellago, C., Hummer, G.: Computing equilibrium free energies using non-equilibrium molecular dynamics. Entropy **16**(1), 41–61 (2014)
56. Morfill, J., Neumann, J., Blank, K., Steinbach, U., Puchner, E.M., Gottschalk, K.E., Gaub, H.E.: Force-based analysis of multidimensional energy landscapes: application of dynamic force spectroscopy and steered molecular dynamics simulations to an antibody fragment-peptide complex. J. Mol. Biol. **381**(5), 1253–1266 (2008)
57. Hummer, G., Szabo, A.: Free energy surfaces from single-molecule force spectroscopy. Acc. Chem. Res. **38**, 504–513 (2005)
58. Suzuki, Y., Dudko, O.K.: Biomolecules under mechanical stress: a simple mechanism of complex behavior. J. Chem. Phys. **134**, 65102 (2011)
59. Bennett, C.H.: Efficient estimation of free energy differences in Monte Carlo data. J. Comp. Phys. **22**(2), 245–268 (1976)
60. Shenfeld, D.K., Xu, H., Eastwood, M.P., Dror, R.O., Shaw, D.E.: Minimizing thermodynamic length to select intermediate states for free-energy calculations and replica-exchange simulations. Phys. Rev. E **80**, 046705 (2009)
61. Kim, S., Kim, Y.W., Talkner, P., Yi, J.: Comparison of free-energy estimators and their dependence on dissipated work. Phys. Rev. E **86**(4), 1–10 (2012)
62. Blaber, S., Sivak, D.A.: Skewed thermodynamic geometry and optimal free energy estimation. J. Chem. Phys. **153**(24), 244119 (2020)
63. Hummer, G., Szabo, A.: Free energy reconstruction from nonequilibrium single-molecule pulling experiments. Proc. Natl. Acad. Sci. U. S. A. **98**(7), 3658–3661 (2001)
64. Andresen, B., Salamon, P.: Thermodynamic geometry determines optimal temperature profile in distillation column. In: Hansen, P., Jacobsen, B., Mosegaard, K. (eds.) Methods and Applications of Inversion, pp. 15–30. Springer, Berlin (2000)
65. Huang, P.S., Boyken, S.E., Baker, D.: The coming of age of de novo protein design. Nature **537**(7620), 320–327 (2016)
66. Schmiedl, T., Seifert, U.: Optimal finite-time processes in stochastic thermodynamics. Phys. Rev. Lett. **98**(10), 1–4 (2007)
67. Crooks, G.E.: Measuring thermodynamic length. Phys. Rev. Lett. **99**, 100602 (2007)

68. Bonança, M.V.S., Deffner, S.: Optimal driving of isothermal processes close to equilibrium. J Chem. Phys. **140**, 244119 (2014)
69. Large, S.J., Sivak, D.A.: Optimal discrete control: minimizing dissipation in discretely driven nonequilibrium systems. J. Stat. Mech. Theory Exp. **2019**(8) (2019)
70. Blaber, S., Louwerse, M.D., Sivak, D.A.: Steps minimize dissipation in rapidly driven stochastic systems. Phys. Rev. E **104**(2), 1–6 (2021)
71. Mandal, D., Jarzynski, C.: Analysis of slow transitions between nonequilibrium steady states. J. Stat. Mech. 063204 (2016)
72. Berkowitz, M., Morgan, J.D., Mccammon, J.A., Northrup, S.H.: Diffusion-controlled reactions: a variational formula for the optimum reaction coordinate. J. Chem. Phys. **79**(10), 5563–2325 (1983)
73. Maragliano, L., Fischer, A., Vanden-Eijnden, E., Ciccotti, G.: String method in collective variables: minimum free energy paths and isocommittor surfaces. J. Chem. Phys. **125**(2), 024106 (2006)
74. Zhao, R., Shen, J., Skeel, R.D.: Maximum flux transition paths of conformational change. J. Chem. Theory Comput. **6**(8), 2411–2423 (2010)
75. Johnson, M.E., Hummer, G.: Characterization of a dynamic string method for the construction of transition pathways in molecular reactions. J. Phys. Chem. B **116**(29), 8573–8583 (2012)
76. Louwerse, M.D., Sivak, D.A.: Information thermodynamics of the transition-path ensemble. Phys. Rev. Lett. **128**, 170602 (2022)
77. Louwerse, M.D., Sivak, D.A.: Multidimensional minimum-work control of a 2D Ising model. J. Chem. Phys. **156**, 194108 (2022)

Chapter 2
Theoretical Background

Abstract In this chapter, some necessary theoretical background is introduced which will inform the work in the remainder of this thesis. Section 2.1 begins with a review of stochastic dynamics, followed by some background into information theory in Sect. 2.1.2 and stochastic thermodynamics in Sect. 2.1.3. The fundamentals of transition-path theory are then reviewed in Sect. 2.2 and the utility of the committor as a reaction coordinate is discussed in more detail. In Sect. 2.3, the thermodynamics of control protocols are reviewed, followed by introduction of minimum-work protocols in the linear-response regime and the generalized friction.

2.1 Stochastic Dynamics and Thermodynamics

Equilibrium thermodynamics explains many observations in the physical sciences, including phase transitions, reaction spontaneity, and reversible processes. As interest in and ability to study nonequilibrium systems grows, integrating system dynamics into descriptions of thermodynamics has become a necessary tool to understand biological phenomena. This motivates the field of stochastic thermodynamics, which provides definitions of work, heat, and entropy for individual trajectories that are then readily applied to nonequilibrium systems. These use stochastic descriptions of system dynamics, where noise provides a random force to the system in addition to conserved and non-conserved forces, for example, that arise from internal energy and nonequilibrium environmental forces.

2.1.1 Ensemble and Trajectory Descriptions of Stochastic Dynamics

Suppose the system is in microstate $\boldsymbol{\phi}$ at time t. In deterministic classical mechanics, $\boldsymbol{\phi}(t)$ describes the position and momentum of each coordinate of the system at time t; by integrating classical equations of motion, the microstate at a later time $\boldsymbol{\phi}(t + \delta t)$ can be determined [1]. Importantly, a trajectory initiated from a point

M. Louwerse, *Efficient Control and Spontaneous Transitions*, Springer Theses,
https://doi.org/10.1007/978-3-031-40534-1_2

$\boldsymbol{\phi}$ follows the same path through phase space each time. This is not the case with stochastic dynamics; each step in the dynamics incorporates random forces, so a trajectory initiated from $\boldsymbol{\phi}$ follows a different path in each trial, leading to a distribution of possible outcomes at a later time $p(\boldsymbol{\phi}, t + \delta t|\boldsymbol{\phi}, t)$. However, with certain assumptions on the random noise, average properties of the ensemble of trajectories can be derived. Formal introductions to stochastic dynamics are presented elsewhere [2]. This section instead focuses on developing some intuition about stochastic dynamics, with relevant equations and notation appearing as necessary.

To begin, assume the system is initialized in state $\boldsymbol{\phi}$ at time t_0 and the system dynamics are simulated via a Markov jump process for n discrete time steps until time t_n to produce a trajectory $\{\boldsymbol{\phi}(t)\} = \{\boldsymbol{\phi}_0, t_0; \boldsymbol{\phi}_1, t_1; \ldots; \boldsymbol{\phi}_n, t_n\}$. It is assumed that the dynamics are Markovian, that is, the next state visited by the system depends only on the current state and not past states visited by the system [2, 3]. The probability of producing a particular trajectory $\{\boldsymbol{\phi}(t)\}$ can then be expressed as

$$p(\{\boldsymbol{\phi}(t)\}) = p(\boldsymbol{\phi}_0, t_0) \prod_{i=1}^{n} p(\boldsymbol{\phi}_i, t_i|\boldsymbol{\phi}_{i-1}, t_{i-1}) , \tag{2.1}$$

where $p(\boldsymbol{\phi}_i, t_i|\boldsymbol{\phi}_{i-1}t_{i-1})$ is the *transition probability*, the probability the system makes a transition from $\boldsymbol{\phi}_{i-1}$ to $\boldsymbol{\phi}_i$ in the time interval $t_{i-1} \rightarrow t_i$, and $p(\boldsymbol{\phi}_0, t_0)$ is the initial probability distribution of the system.

The transition probabilities should be chosen such that at long times the system distribution reproduces the distribution of interest. To generate the dynamics of an equilibrium distribution $\pi(\boldsymbol{\phi})$, the transition probabilities should be independent of time and obey the principle of *detailed-balance* [2]

$$p(\boldsymbol{\phi}|\boldsymbol{\phi}')\pi(\boldsymbol{\phi}') = p(\boldsymbol{\phi}'|\boldsymbol{\phi})\pi(\boldsymbol{\phi}) . \tag{2.2}$$

For each pair of microstates, the equilibrium probability the system is in $\boldsymbol{\phi}'$ and makes a transition to $\boldsymbol{\phi}$ is equal to the probability the system is in $\boldsymbol{\phi}$ and makes a transition to $\boldsymbol{\phi}'$. Rearranging, the ratio of transition rates is equal to the ratio of equilibrium probabilities

$$\frac{p(\boldsymbol{\phi}|\boldsymbol{\phi}')}{p(\boldsymbol{\phi}'|\boldsymbol{\phi})} = \frac{\pi(\boldsymbol{\phi})}{\pi(\boldsymbol{\phi}')} . \tag{2.3}$$

There are many possible ways to choose transition probabilities that satisfy these conditions. Given an initial state $\boldsymbol{\phi}_0$ drawn from the initial probability distribution $p(\boldsymbol{\phi}_0, t_0)$, a general Monte Carlo procedure generates a new trial state $\boldsymbol{\phi}^*$ with probability $p^{\text{trial}}(\boldsymbol{\phi}^*|\boldsymbol{\phi}_0)$ and then draws a random number between 0 and 1 to determine whether or not the new state is accepted. The total transition probability from $\boldsymbol{\phi}_0$ to $\boldsymbol{\phi}^*$,

$$p(\boldsymbol{\phi}_1 = \boldsymbol{\phi}^*, t_1|\boldsymbol{\phi}_0, t_0) = p^{\text{trial}}(\boldsymbol{\phi}^*|\boldsymbol{\phi}_0)p^{\text{accept}}(\boldsymbol{\phi}^*|\boldsymbol{\phi}_0) , \tag{2.4}$$

where $p^{\text{accept}}(\boldsymbol{\phi}^*|\boldsymbol{\phi}_0)$ is the acceptance probability that must be chosen to ensure dynamics obey detailed balance. If the random number is less than the transition probability, the system moves to $\boldsymbol{\phi}_1 = \boldsymbol{\phi}^*$ at time t_1, otherwise it stays in state $\boldsymbol{\phi}_0$ at time t_1. The process is repeated till a trajectory of the desired length is generated.

There are many choices for generating the trial state, depending on the system of interest. In molecular simulations, random perturbations to positions and momenta of atoms could be applied [4]. For the Ising model studied in this thesis (introduced in Sect. 4.2), a trial state is generated by randomly choosing one of N spins making up the system from a uniform distribution and flipping it, thus $p^{\text{trial}}(\boldsymbol{\phi}^*|\boldsymbol{\phi}_0) = N^{-1}$. If the trial probability is symmetric, $p^{\text{trial}}(\boldsymbol{\phi}^*|\boldsymbol{\phi}_0) = p^{\text{trial}}(\boldsymbol{\phi}_0|\boldsymbol{\phi}_*)$, the constraint on transition probabilities imposed by detailed balance falls on the acceptance probability. For physical systems with energy $V(\boldsymbol{\phi})$ that relax to a Boltzmann distribution, the acceptance probabilities must satisfy

$$\frac{p^{\text{accept}}(\boldsymbol{\phi}^*|\boldsymbol{\phi}_0)}{p^{\text{accept}}(\boldsymbol{\phi}_0|\boldsymbol{\phi}^*)} = e^{-\beta[V(\boldsymbol{\phi}^*)-V(\boldsymbol{\phi}_0)]} . \tag{2.5}$$

Common choices for the acceptance probability include the Metropolis criterion [5] and the Glauber criterion [6], but other choices are possible as long as they obey certain constraints to ensure the process converges to the desired distribution. The Glauber acceptance probability is used in this thesis, which has the form

$$p^{\text{accept,GL}}(\boldsymbol{\phi}^*|\boldsymbol{\phi}_0) = \frac{1}{1+e^{\beta[V(\boldsymbol{\phi}^*)-V(\boldsymbol{\phi}_0)]}} . \tag{2.6}$$

The Glauber acceptance criterion (2.6) satisfies detailed balance,

$$\frac{p^{\text{accept,GL}}(\boldsymbol{\phi}^*|\boldsymbol{\phi}_0)}{p^{\text{accept,GL}}(\boldsymbol{\phi}_0|\boldsymbol{\phi}^*)} = \frac{1+e^{\beta[V(\boldsymbol{\phi}_0)-V(\boldsymbol{\phi}^*)]}}{1+e^{\beta[V(\boldsymbol{\phi}^*)-V(\boldsymbol{\phi}_0)]}} \tag{2.7a}$$

$$= \frac{e^{\beta V(\boldsymbol{\phi}^*)}+e^{\beta V(\boldsymbol{\phi}_0)}}{e^{\beta V(\boldsymbol{\phi}_0)}+e^{\beta V(\boldsymbol{\phi}^*)}} \frac{e^{\beta V(\boldsymbol{\phi}_0)}}{e^{\beta V(\boldsymbol{\phi}^*)}} \tag{2.7b}$$

$$= e^{-\beta[V(\boldsymbol{\phi}^*)-V(\boldsymbol{\phi}_0)]} . \tag{2.7c}$$

Equation (2.1) is the probability for a single trajectory generated by a Markov jump process. It is often of interest to obtain average quantities over all possible trajectories $\{\boldsymbol{\phi}(t)\}$; it is therefore useful to have a description of the way the system's probability distribution evolves in time $p(\boldsymbol{\phi}, t)$ from an initial distribution. This is commonly expressed as a Master equation [2]

$$\mathrm{d}_t p(\boldsymbol{\phi}, t) = \sum_{\boldsymbol{\phi}'} T_{\boldsymbol{\phi}\boldsymbol{\phi}'}(t) p(\boldsymbol{\phi}', t) , \tag{2.8}$$

where $T_{\boldsymbol{\phi}\boldsymbol{\phi}'}(t) = \lim_{\delta t \to 0} p(\boldsymbol{\phi}, t + \delta t | \boldsymbol{\phi}', t)/\delta t$ is the *transition rate* of $\boldsymbol{\phi}' \to \boldsymbol{\phi}$ transition at time t, which is expressed as the transition probability per infinitesimal time step divided by time step δt. The $\boldsymbol{\phi} \to \boldsymbol{\phi}$ transition rate is defined as

$$T_{\boldsymbol{\phi}\boldsymbol{\phi}} = -\sum_{\boldsymbol{\phi}' \neq \boldsymbol{\phi}} T_{\boldsymbol{\phi}'\boldsymbol{\phi}} \tag{2.9}$$

to ensure the probability distribution remains normalized.

The Master equation describes how the probability distribution changes from the initial condition $p(\boldsymbol{\phi}_0, t_0)$ as the system evolves stochastically. When the transition rate is independent of time, the system may relax to a stationary distribution $p^{\text{ss}}(\boldsymbol{\phi})$

$$0 = \mathrm{d}_t p^{\text{ss}}(\boldsymbol{\phi}) = \sum_{\boldsymbol{\phi}'} T_{\boldsymbol{\phi}\boldsymbol{\phi}'} p^{\text{ss}}(\boldsymbol{\phi}')\,, \tag{2.10}$$

where the probability of each state is constant. This is the global balance condition [2], ensuring that there are no changes to the probability of any state $\boldsymbol{\phi}$

$$0 = \sum_{\boldsymbol{\phi}'} \left[T_{\boldsymbol{\phi}'\boldsymbol{\phi}} p^{\text{ss}}(\boldsymbol{\phi}) - T_{\boldsymbol{\phi}\boldsymbol{\phi}'} p^{\text{ss}}(\boldsymbol{\phi}') \right] . \tag{2.11}$$

However, the stationary distribution is not necessarily an equilibrium distribution; there may be net probability flows in the system that are balanced such that there is no overall change in the probability of each state. Equilibrium requires the stronger condition of detailed balance, which can be expressed in terms of transition rates as

$$T_{\boldsymbol{\phi}\boldsymbol{\phi}'} \pi(\boldsymbol{\phi}') = T_{\boldsymbol{\phi}'\boldsymbol{\phi}} \pi(\boldsymbol{\phi})\,, \tag{2.12}$$

for all pairs of states $\boldsymbol{\phi}, \boldsymbol{\phi}'$. Not only is the overall probability in each state conserved at equilibrium, but probability flux between all states is equal in both directions, resulting in no flux or probability flows throughout the state space. If global balance holds but detailed balance does not, the stationary distribution can be called a *nonequilibrium steady state* [3].

2.1.2 Entropy and Information Theory

Statistical mechanics treats the system state $\boldsymbol{\phi}$ as a random variable $\boldsymbol{\Phi}$. This allows an information-theoretic perspective on entropy that parallels its thermodynamic definition. The foundations of information theory were laid out by Claude Shannon at Bell Labs in 1948 [7]. Since then, it has been recognized that information is physical, as elegantly argued by Landauer [8]. Some key concepts from information theory are briefly reviewed to better elucidate this connection, which will be relevant for Chap. 3.

First, consider quantifying the uncertainty about a random variable X before taking a measurement. Throughout, the convention of capital symbols to denote a random variable and lowercase symbols to denote a particular realization of the random variable is used. The uncertainty about discrete random variable X with probability mass function $p(X = x)$ is given by its Shannon entropy, defined to be

$$H(X) \equiv -\sum_{x} p(x) \ln p(x) . \tag{2.13}$$

Notably, this is the same form for thermodynamic entropy derived by Gibbs, up to a constant k_B [9]. The average amount of information gained about the distribution upon measurement of the random variable is equal to its Shannon entropy, therefore information represents a reduction in uncertainty.

Information theory gains more power by considering multiple random variables X and Y. Consider X and Y jointly distributed according to the probability mass function $p(X = x, Y = y)$. For example, X and Y could be coordinates representing the state of a two-dimensional system. The joint entropy of the joint random variable (X, Y) is

$$H(X, Y) \equiv -\sum_{x,y} p(x, y) \ln p(x, y) . \tag{2.14}$$

Similarly, the marginal entropy of one of these random variables (without loss of generality X) is expressed in terms of its marginal distribution $p(x) = \sum_y p(x, y)$ as

$$H(X) = -\sum_{x} p(x) \ln p(x) \tag{2.15a}$$

$$= -\sum_{x,y} p(x, y) \ln \sum_{y} p(x, y) . \tag{2.15b}$$

This is the uncertainty about the joint state that can be reduced from observing only X. The conditional entropy of X given Y can be similarly defined as

$$H(X|Y) \equiv -\sum_{x,y} p(x, y) \ln p(x|y) , \tag{2.16}$$

which represents the remaining uncertainty about X when Y is known. This hints that information is shared between X and Y, which can be precisely quantified by the mutual information between X and Y,

$$I(X; Y) \equiv \sum_{x,y} p(x, y) \ln \frac{p(x, y)}{p(x)p(y)} . \tag{2.17}$$

Mutual information is a nonlinear statistical measure of the relationship between random variables X and Y, specifically measuring the reduction in uncertainty about one variable from measuring the other,

$$I(X;Y) = H(X) - H(X|Y) \tag{2.18a}$$

$$= H(Y) - H(Y|X) \tag{2.18b}$$

$$= I(Y;X)\,. \tag{2.18c}$$

It is a symmetric quantity that is positive semi-definite

$$I(X;Y) = -\sum_{x,y} p(x,y) \ln \frac{p(x)p(y)}{p(x,y)} \tag{2.19a}$$

$$\geq \ln \sum_{x,y} p(x,y) \frac{p(x)p(y)}{p(x,y)} \tag{2.19b}$$

$$= \ln \sum_{x,y} p(x)p(y) \tag{2.19c}$$

$$= \ln 1 \tag{2.19d}$$

$$= 0\,, \tag{2.19e}$$

and zero if and only if X and Y are independent $[p(x,y) = p(x)p(y)]$. Equation (2.19b) uses Jensen's inequality [7].

Information theory also has broad relevance throughout the field of stochastic thermodynamics, as an essential aspect of describing many classes of nonequilibrium systems [10–12].

2.1.3 Stochastic Thermodynamics

Stochastic thermodynamics defines thermodynamic quantities for individual trajectories. Here, a trajectory generated by a Markov jump process introduced in Sect. 2.1.1 is considered. The transition rate for a $\boldsymbol{\phi}' \to \boldsymbol{\phi}$ transition is $T_{\boldsymbol{\phi}\boldsymbol{\phi}'}$, and the probability the system is in state $\boldsymbol{\phi}$ at time t is $p(\boldsymbol{\phi}, t)$. Let $V(\boldsymbol{\phi}, t)$ denote the energy of state $\boldsymbol{\phi}$ at time t. The first law of thermodynamics accounts for changes to the system's energy as work or heat,

$$\mathrm{d}_t V(t) = \sum_{\boldsymbol{\phi}} \mathrm{d}_t \left[p(\boldsymbol{\phi}, t) V(\boldsymbol{\phi}, t)\right] \tag{2.20a}$$

$$= \sum_{\boldsymbol{\phi}} \left[V(\boldsymbol{\phi}, t) \mathrm{d}_t p(\boldsymbol{\phi}, t) + p(\boldsymbol{\phi}, t) \mathrm{d}_t V(\boldsymbol{\phi}, t)\right] \tag{2.20b}$$

$$= \dot{Q}(t) + \dot{W}(t) \tag{2.20c}$$

where the overdot notation indicates a rate (here, of heat and work) that cannot be expressed as the time derivative of a state function. Work is performed by an external agent making changes to the energy of microstates, while heat is produced during system transitions by exchange of energy with thermodynamic reservoirs. The sign convention is used that positive heat and work correspond to energy flows into the system. Further discussion of work done on the system during a control protocol is discussed in Sect. 2.3.2.

Heat flow to the system for the instantaneous transition from $\boldsymbol{\phi}' \to \boldsymbol{\phi}$ is the change in system energy during the transition at time t, which is equal to a log ratio of transition rates that obey detailed balance [13],

$$q_{\boldsymbol{\phi}\boldsymbol{\phi}'}(t) \equiv k_{\mathrm{B}}T[V(\boldsymbol{\phi}, t) - V(\boldsymbol{\phi}', t)] = k_{\mathrm{B}}T \ln \frac{T_{\boldsymbol{\phi}'\boldsymbol{\phi}}(t)}{T_{\boldsymbol{\phi}\boldsymbol{\phi}'}(t)} . \tag{2.21}$$

Assuming the reservoirs in contact with the system remain in equilibrium, the heat flow is proportional to the negative entropy change in the environment

$$q_{\boldsymbol{\phi}\boldsymbol{\phi}'}(t) = -k_{\mathrm{B}}T\, h^{\mathrm{env}}_{\boldsymbol{\phi}\boldsymbol{\phi}'}(t) . \tag{2.22}$$

Entropy is expressed as a dimensionless quantity throughout this thesis. The change in system entropy for a $\boldsymbol{\phi}' \to \boldsymbol{\phi}$ transition is the difference in specific entropy $h(\boldsymbol{\phi}, t) \equiv -\ln p(\boldsymbol{\phi}, t)$,

$$\Delta h^{\mathrm{sys}}_{\boldsymbol{\phi}\boldsymbol{\phi}'}(t) \equiv \ln \frac{p(\boldsymbol{\phi}', t)}{p(\boldsymbol{\phi}, t)} . \tag{2.23}$$

Thus the total (irreversible) entropy production for a $\boldsymbol{\phi}' \to \boldsymbol{\phi}$ transition is

$$h^{\mathrm{irr}}_{\boldsymbol{\phi}\boldsymbol{\phi}'}(t) = \Delta h^{\mathrm{sys}}_{\boldsymbol{\phi}\boldsymbol{\phi}'}(t) + h^{\mathrm{env}}_{\boldsymbol{\phi}\boldsymbol{\phi}'}(t) \tag{2.24a}$$

$$= \ln \frac{T_{\boldsymbol{\phi}\boldsymbol{\phi}'}(t)p(\boldsymbol{\phi}', t)}{T_{\boldsymbol{\phi}'\boldsymbol{\phi}}(t)p(\boldsymbol{\phi}, t)} , \tag{2.24b}$$

the log ratio of the forward and reverse flux for a transition between $\boldsymbol{\phi}'$ and $\boldsymbol{\phi}$ at time t, also called the thermodynamic force [13]. If the system is in equilibrium so there is equal flux of forward and reverse transitions $[T_{\boldsymbol{\phi}\boldsymbol{\phi}'}\pi(\boldsymbol{\phi}') = T_{\boldsymbol{\phi}'\boldsymbol{\phi}}\pi(\boldsymbol{\phi})]$, the total entropy production is zero for each transition,

$$h^{\mathrm{irr}}_{\boldsymbol{\phi}\boldsymbol{\phi}'} = h^{\mathrm{sys}}_{\boldsymbol{\phi}\boldsymbol{\phi}'} + h^{\mathrm{env}}_{\boldsymbol{\phi}\boldsymbol{\phi}'} \tag{2.25a}$$

$$= -k_{\mathrm{B}}T \ln \frac{T_{\boldsymbol{\phi}'\boldsymbol{\phi}}}{T_{\boldsymbol{\phi}\boldsymbol{\phi}'}} + k_{\mathrm{B}}T \ln \frac{\pi(\boldsymbol{\phi}')}{\pi(\boldsymbol{\phi})} \tag{2.25b}$$

$$= 0. \tag{2.25c}$$

Assuming the system has reached steady state $p^{\mathrm{ss}}(\boldsymbol{\phi})$ with its reservoirs (which may be out of equilibrium), multiplying the thermodynamic quantities defined for each transition by the net probability flux of the transition

$$J_{\boldsymbol{\phi}\boldsymbol{\phi}'} \equiv T_{\boldsymbol{\phi}\boldsymbol{\phi}'} p^{\mathrm{ss}}(\boldsymbol{\phi}') - T_{\boldsymbol{\phi}'\boldsymbol{\phi}} p^{\mathrm{ss}}(\boldsymbol{\phi}) , \tag{2.26}$$

gives ensemble-averaged thermodynamic quantities. The steady-state heat flow to the system is

$$\dot{Q} = \frac{k_{\mathrm{B}}T}{2} \sum_{\boldsymbol{\phi}\boldsymbol{\phi}'} J_{\boldsymbol{\phi}\boldsymbol{\phi}'} \ln \frac{T_{\boldsymbol{\phi}'\boldsymbol{\phi}}}{T_{\boldsymbol{\phi}\boldsymbol{\phi}'}} \tag{2.27a}$$

$$= -k_{\mathrm{B}}T \dot{H}^{\mathrm{env}} , \tag{2.27b}$$

where the factor of $\frac{1}{2}$ accounts for the inclusion of each pair of states twice in the sum.

System entropy is unchanging at steady state (even for nonequilibrium steady states) since it is a state function of an unchanging distribution:

$$\mathrm{d}_t H^{\mathrm{sys}} = \frac{1}{2} \sum_{\boldsymbol{\phi}\boldsymbol{\phi}'} J_{\boldsymbol{\phi}\boldsymbol{\phi}'} \ln \frac{p^{\mathrm{ss}}(\boldsymbol{\phi}')}{p^{\mathrm{ss}}(\boldsymbol{\phi})} \tag{2.28a}$$

$$= \frac{1}{2} \sum_{\boldsymbol{\phi}\boldsymbol{\phi}'} J_{\boldsymbol{\phi}\boldsymbol{\phi}'} \left[\ln p^{\mathrm{ss}}(\boldsymbol{\phi}') - \ln p^{\mathrm{ss}}(\boldsymbol{\phi})\right] \tag{2.28b}$$

$$= \frac{1}{2} \left[\sum_{\boldsymbol{\phi}'} \ln p^{\mathrm{ss}}(\boldsymbol{\phi}') \sum_{\boldsymbol{\phi}} J_{\boldsymbol{\phi}\boldsymbol{\phi}'} - \sum_{\boldsymbol{\phi}} \ln p^{\mathrm{ss}}(\boldsymbol{\phi}) \sum_{\boldsymbol{\phi}'} J_{\boldsymbol{\phi}\boldsymbol{\phi}'} \right] \tag{2.28c}$$

$$= 0 , \tag{2.28d}$$

using $\sum_{\boldsymbol{\phi}'} J_{\boldsymbol{\phi}\boldsymbol{\phi}'} = \sum_{\boldsymbol{\phi}'} J_{\boldsymbol{\phi}'\boldsymbol{\phi}} = 0$ for all $\boldsymbol{\phi}$ when the system is at steady state (Eq. (2.11)). Therefore the total entropy production rate is equal to the rate of change in environmental entropy

$$\dot{H}^{\mathrm{irr}} = \sum_{\boldsymbol{\phi}>\boldsymbol{\phi}'} J_{\boldsymbol{\phi}\boldsymbol{\phi}'} \ln \frac{T_{\boldsymbol{\phi}\boldsymbol{\phi}'} p^{\mathrm{ss}}(\boldsymbol{\phi}')}{T_{\boldsymbol{\phi}'\boldsymbol{\phi}} p^{\mathrm{ss}}(\boldsymbol{\phi})} \tag{2.29a}$$

$$= \sum_{\boldsymbol{\phi}\boldsymbol{\phi}'} J_{\boldsymbol{\phi}\boldsymbol{\phi}'} \ln \frac{p^{\mathrm{ss}}(\boldsymbol{\phi}')}{p^{\mathrm{ss}}(\boldsymbol{\phi})} + \sum_{\boldsymbol{\phi}\boldsymbol{\phi}'} J_{\boldsymbol{\phi}\boldsymbol{\phi}'} \ln \frac{T_{\boldsymbol{\phi}\boldsymbol{\phi}'}}{T_{\boldsymbol{\phi}'\boldsymbol{\phi}}} \tag{2.29b}$$

$$= \mathrm{d}_t H^{\mathrm{sys}} + \dot{H}^{\mathrm{env}} \tag{2.29c}$$

$$= 0 + \dot{H}^{\mathrm{env}} \geq 0 , \tag{2.29d}$$

and is positive semi-definite. $\dot{H}^{\text{irr}} = 0$ if and only if detailed balance (2.12) holds. Irreversible entropy production is the driving force behind all spontaneous processes, and there is significant interest in understanding its role in biological processes [14, 15]. Chapter 3 shows that irreversible entropy production also plays a role in characterizing reaction mechanisms.

2.2 Theory and Analysis of Spontaneous Transition Paths

Given the stochastic dynamics described in Sect. 2.1.1, the theory of transition paths can now be introduced. Transition-path theory for Markov jump processes in discrete state spaces [16, 17] is presented, but the theory was originally developed for continuous systems [18]. Concepts of the transition-path ensemble, the committor, and reaction statistics computed from the committor are reviewed below.

2.2.1 Transition-Path Theory

Assume a system's dynamics are governed by the Master equation Eq. (2.8) with $\boldsymbol{\phi}' \to \boldsymbol{\phi}$ transition rate $T_{\boldsymbol{\phi}\boldsymbol{\phi}'}$ and that the system has reached the equilibrium distribution $\pi(\boldsymbol{\phi})$ such that detailed-balance holds (Eq. (2.12)). Select from the state space $\boldsymbol{\Phi}$ two distinct subsets of reactant microstates A and product microstates B. These subsets should be non-overlapping so the system must travel through an intermediate region I as it progresses from A to B. During a long ergodic equilibrium *supertrajectory*, the system transitions between A and B infinitely many times; the transition-path ensemble (TPE) is the set of trajectory segments (also called *reactive trajectories* or *transition paths*) that travel directly from A to B. Statistics such as reactive flux, mean first-passage time, and mean transition-path duration are derived from the frequency of $A \to B$ transition events using transition-path theory.

The joint probability that the system is in state $\boldsymbol{\phi}$ and on a reactive trajectory segment

$$p(\boldsymbol{\phi}, A \to B) = \pi(\boldsymbol{\phi}) p(A \to B|\boldsymbol{\phi}) , \tag{2.30}$$

is conveniently expressed as the product of the marginal (equilibrium) probability the system is in state $\boldsymbol{\phi}$ and the conditional probability $p(A \to B|\boldsymbol{\phi})$ that the system is currently on an $A \to B$ reactive trajectory given that it is currently in state $\boldsymbol{\phi}$. It then remains to find this conditional probability.

Define the *forward committor* $q^+(\boldsymbol{\phi})$ to be the probability the system will next reach B given that it is in state $\boldsymbol{\phi}$, and the *backward committor* $q^-(\boldsymbol{\phi})$ to be the probability the system was more recently in A than B given that it is currently in state $\boldsymbol{\phi}$. These committor functions can be defined recursively for discrete-state systems:

$$\sum_{\boldsymbol{\phi}'} T_{\boldsymbol{\phi}'\boldsymbol{\phi}} q^+(\boldsymbol{\phi}') = 0 \tag{2.31a}$$

$$\sum_{\boldsymbol{\phi}'} T_{\boldsymbol{\phi}'\boldsymbol{\phi}} \frac{\pi(\boldsymbol{\phi})}{\pi(\boldsymbol{\phi}')} q^-(\boldsymbol{\phi}') = 0 \tag{2.31b}$$

with boundary conditions $q^+(\boldsymbol{\phi}) = 0(1)$ and $q^-(\boldsymbol{\phi}) = 1(0)$ for $\boldsymbol{\phi} \in A(B)$. For discrete-state systems, the committors can be calculated numerically by solving the system of linear equations (2.31) to find the committor of each state.

For equilibrium systems, the principle of microscopic reversibility [2] is used to relate the backward committor to the forward committor. The backward committor $q^-(\boldsymbol{\phi})$ is the probability the system in state $\boldsymbol{\phi}$ was generated from a trajectory that originated in A. Each such trajectory from A to $\boldsymbol{\phi}$ has an equally probable time-reversed trajectory from $\boldsymbol{\phi}$ to A. Therefore the ratio of $A \rightarrow \boldsymbol{\phi}$ vs. $B \rightarrow \boldsymbol{\phi}$ trajectories is equal to the ratio of $\boldsymbol{\phi} \rightarrow A$ vs. $\boldsymbol{\phi} \rightarrow B$ trajectories. Since the probability of a $\boldsymbol{\phi} \rightarrow A$ trajectory is $1-q^+(\boldsymbol{\phi})$, the forward and backward committor functions are related by the equality

$$1 - q^+(\boldsymbol{\phi}) = q^-(\boldsymbol{\phi}) \,. \tag{2.32}$$

This implies that due to time-reversal symmetry of equilibrium dynamics, either committor function (usually the forward committor is adopted) is sufficient on its own to fully describe the statistics of reactive trajectories in the intermediate state space between A and B.

With these two definitions, the joint probability the system is in state $\boldsymbol{\phi}$ and on a reactive trajectory is expressed as

$$p(\boldsymbol{\phi}, A \rightarrow B) = \pi(\boldsymbol{\phi}) q^-(\boldsymbol{\phi}) q^+(\boldsymbol{\phi}) \,. \tag{2.33}$$

In words, this is the equilibrium probability the system is in state $\boldsymbol{\phi}$ and was more recently in A and next reaches B. The independence (on the rhs) of the probability the system was more recently in A given it is currently in $\boldsymbol{\phi}$ and the probability the system next reaches B conditioned on state $\boldsymbol{\phi}$ can be understood from the Markov property: whether the system was more recently in A or B does not influence future dynamics since they are assumed to be memoryless and depend only on the current state $\boldsymbol{\phi}$. Therefore the probability the system next proceeds to B (rather than A), conditioned on current state $\boldsymbol{\phi}$, is independent of whether the system was more recently in A or B. Equation (2.33) readily yields the probability distribution of reactive trajectories,

$$p(\boldsymbol{\phi}|A \rightarrow B) = \frac{\pi(\boldsymbol{\phi}) q^-(\boldsymbol{\phi}) q^+(\boldsymbol{\phi})}{p_{A\rightarrow B}} \,, \tag{2.34}$$

where

$$p_{A\to B} = \sum_{\boldsymbol{\phi}} \pi(\boldsymbol{\phi}) q^-(\boldsymbol{\phi}) q^+(\boldsymbol{\phi}) \tag{2.35}$$

is the probability the system is on an $A \to B$ reactive trajectory at time t during the ergodic supertrajectory.

A particularly useful output of transition-path theory is a description of the probability current of reactive trajectories. Consider the transition-path ensemble to be in a steady state where the system is initiated at the boundary of A and is constrained to reach B without returning to A. The system is instantly regenerated at the boundary of A once it reaches the boundary of B. This results in a steady-state current of trajectories flowing from A to B, with reactive probability current between states $\boldsymbol{\phi}'$ and $\boldsymbol{\phi}$ given by

$$J_{\boldsymbol{\phi}\boldsymbol{\phi}'}^{A\to B} = \pi(\boldsymbol{\phi}') q^-(\boldsymbol{\phi}') T_{\boldsymbol{\phi}\boldsymbol{\phi}'} q^+(\boldsymbol{\phi}) \, . \tag{2.36}$$

This is the joint probability that the system is in state $\boldsymbol{\phi}'$, was more recently in A than B, makes a transition in simulation time $\mathrm{d}t$ to state $\boldsymbol{\phi}$, and will next reach B before A. When averaged over all transitions that leave A and start a transition path, or all transitions that finish transition paths with a jump into B, the reactive probability current yields the flux of $A \to B$ transitions through the intermediate region I

$$\nu_{A\to B} = \sum_{\boldsymbol{\phi}'\in A, \boldsymbol{\phi}\in I} J_{\boldsymbol{\phi}\boldsymbol{\phi}'}^{A\to B} \tag{2.37a}$$

$$= \sum_{\boldsymbol{\phi}'\in I, \boldsymbol{\phi}\in B} J_{\boldsymbol{\phi}\boldsymbol{\phi}'}^{A\to B} \, . \tag{2.37b}$$

In fact, any dividing surface between A and B can be used to find the reaction rate, since the steady-state reactive flux is the same through any dividing surface. Additionally, the flux of $A \to B$ transitions is equal to the flux of $B \to A$ "reverse" transitions, since the system is at equilibrium. The reactive flux is related to the rate constants by

$$k_{A\to B} = \frac{\nu_{A\to B}}{p_A} \tag{2.38a}$$

$$k_{B\to A} = \frac{\nu_{A\to B}}{p_B} \tag{2.38b}$$

where

$$p_A = \sum_{\boldsymbol{\phi}} \pi(\boldsymbol{\phi}) q^-(\boldsymbol{\phi}) \tag{2.39a}$$

$$p_B = \sum_{\boldsymbol{\phi}} \pi(\boldsymbol{\phi})[1 - q^-(\boldsymbol{\phi})] \tag{2.39b}$$

is the probability the system was more recently in A (B) than in B (A) at some time t during the ergodic supertrajectory. The mean first-passage time to B (A) given that the system is currently in A (B) is the inverse of the rate constant

$$\tau^{\text{FP}}_{A\to B} = k^{-1}_{A\to B} \tag{2.40a}$$

$$\tau^{\text{FP}}_{B\to A} = k^{-1}_{B\to A} \,. \tag{2.40b}$$

Another quantity available from transition-path theory is the mean duration of transition paths $\tau_{A\to B}$, which is derived from the probability $p_{A\to B}$ that the system is currently on an $A \to B$ transition path and the mean round-trip time [19]

$$\tau^{\text{RT}} \equiv \tau^{\text{FP}}_{A\to B} + \tau^{\text{FP}}_{B\to A} \tag{2.41}$$

for the system starting in A to reach B and come back to A for the first time,

$$\tau_{A\to B} = p_{A\to B}\, \tau^{\text{RT}} \tag{2.42a}$$

$$= p_{A\to B} \left(\tau^{\text{FP}}_{A\to B} + \tau^{\text{FP}}_{B\to A}\right) \tag{2.42b}$$

$$= p_{A\to B} \left(\frac{1}{k_{A\to B}} + \frac{1}{k_{B\to A}}\right) \tag{2.42c}$$

$$= p_{A\to B} \left(\frac{p_A}{\nu_{A\to B}} + \frac{p_B}{\nu_{A\to B}}\right) \tag{2.42d}$$

$$= \frac{p_{A\to B}}{\nu_{A\to B}} \,, \tag{2.42e}$$

using

$$p_A + p_B = \sum_{\boldsymbol{\phi}} \pi(\boldsymbol{\phi}) q^-(\boldsymbol{\phi}) + \sum_{\boldsymbol{\phi}} \pi(\boldsymbol{\phi})[1 - q^-(\boldsymbol{\phi})] \tag{2.43a}$$

$$= \sum_{\boldsymbol{\phi}} \pi(\boldsymbol{\phi}) \tag{2.43b}$$

$$= 1 \tag{2.43c}$$

to obtain (2.42e). Analogous expressions have been derived for continuous state-space systems [18].

Transition-path theory has become the leading tool for describing reaction events, allowing calculation of average transition-path properties using the committor. Additionally, the committor is widely considered the "true" or "optimal" reaction coordinate [20–22]. Importantly, it provides language to describe reactions where an intuitive system coordinate that acts as a reaction coordinate cannot necessarily be found, and it provides quantitative information about the rate of the reaction through the reactive flux.

The committor and transition-path theory can also provide mechanistic information by highlighting regions of conformational space that are particularly relevant to the reaction [23]. The committor foliates the conformational space between A and B into a series of isocommittor surfaces (all microstates with the same value of the committor), and important features of the system on these surfaces can be determined. For example, the reactive probability distribution may lie within a "transition tube" in an appropriately chosen collective-variable space, so that the mechanism can be described by a single *most likely transition path* that in some way averages the distribution on each isocommittor surfaces. A minimum-free energy path can be found that is parallel to the free-energy gradient in collective-variable space (analogous to the minimum-energy path in the full configuration space) [23–25], or a maximum-flux transition path [26, 27] that maximizes the reactive flux on each isocommittor surface. Algorithms to compute these paths using molecular dynamics simulations in a space of many collective variables have been developed [25, 28]. Importantly, they require that the committor in chosen collective-variable space is a good approximation to the committor in the full configuration space.

2.3 Minimum-Work Control Theory

In this section, the thermodynamic foundation necessary to understand minimum-work control protocols in the linear-response regime is discussed, beginning with a discussion of control-parameter space and its relationship to the conformational space it biases.

2.3.1 Control Parameters and Collective Variables

As introduced in Sect. 1.2.1, there are many choices for control parameters to drive a system, such as temperature, pressure, and external magnetic fields. Since it is of interest to drive the system between metastable states, control parameters with clear correspondence to system collective variables are chosen. For example, a collective variable can be chosen that is presumed to have relevance to the reaction, then the system can be restrained to particular values of this collective variable using a harmonic trap. The centre of the trap (and/or its spring constant) acts as a control parameter, and the system is driven between values of the chosen collective variable corresponding to metastable states by a control protocol that moves the center of the trap [29–31].

The control parameter biases the system's internal energy $E_{\mathrm{int}}(\boldsymbol{\phi})$ with a control-parameter-dependent external energy $E_{\mathrm{ext}}(\boldsymbol{\phi}, \boldsymbol{\lambda})$, to give total energy

$$V(\boldsymbol{\phi}, \boldsymbol{\lambda}) = E_{\mathrm{int}}(\boldsymbol{\phi}) + E_{\mathrm{ext}}(\boldsymbol{\phi}, \boldsymbol{\lambda}) \,. \tag{2.44}$$

The total potential is related to the equilibrium probability distribution of the system at each control parameter,

$$\pi(\boldsymbol{\phi}|\boldsymbol{\lambda}) = e^{\beta[F^{\mathrm{eq}}(\boldsymbol{\lambda}) - V(\boldsymbol{\phi},\boldsymbol{\lambda})]} \,, \tag{2.45}$$

with equilibrium free energy given $\boldsymbol{\lambda}$ of

$$F^{\mathrm{eq}}(\boldsymbol{\lambda}) = -k_{\mathrm{B}}T \ln \sum_{\boldsymbol{\phi}} e^{-\beta V(\boldsymbol{\phi},\boldsymbol{\lambda})} \tag{2.46a}$$

$$= \langle V(\boldsymbol{\phi}, \boldsymbol{\lambda})\rangle_{\boldsymbol{\lambda}}^{\mathrm{eq}} - k_{\mathrm{B}}T H^{\mathrm{eq}}(\boldsymbol{\lambda}) \,, \tag{2.46b}$$

where $\langle \ldots \rangle_{\boldsymbol{\lambda}}^{\mathrm{eq}}$ denotes an average over the equilibrium distribution $\pi(\boldsymbol{\phi}|\boldsymbol{\lambda})$, and

$$H^{\mathrm{eq}}(\boldsymbol{\lambda}) = -\sum_{\boldsymbol{\phi}} \pi(\boldsymbol{\phi}|\boldsymbol{\lambda}) \ln \pi(\boldsymbol{\phi}|\boldsymbol{\lambda}) \tag{2.47}$$

is the entropy of the equilibrium distribution at $\boldsymbol{\lambda}$. Appropriately chosen control parameters allow the system to preferentially sample regions of conformational space that are typically inaccessible to the system but may be visited transiently during a reaction.

Since the conformational space is typically too high-dimensional to interpret, sampling is often done in a lower-dimensional space defined by a set of collective variables $\boldsymbol{X}(\boldsymbol{\phi})$ to identify metastable conformations and free-energy barriers between them. A single point $\boldsymbol{x} = \boldsymbol{X}(\boldsymbol{\phi})$ in collective-variable space corresponds to a manifold of microstates in the higher-dimensional conformational space, so calculation of the marginal equilibrium probability of these coarse variables $\pi(\boldsymbol{X}(\boldsymbol{\phi}))$ in the absence of control ($E_{\mathrm{ext}}(\boldsymbol{\phi}, \boldsymbol{\lambda}) = 0$) is desired. This is expressed in terms of the free energy of a particular value of collective variable $\boldsymbol{x}$,

$$F(\boldsymbol{x}) = -k_{\mathrm{B}}T \ln \sum_{\boldsymbol{\phi}} e^{-\beta E_{\mathrm{int}}(\boldsymbol{\phi})} \delta\left[\boldsymbol{X}(\boldsymbol{\phi}) - \boldsymbol{x}\right] \tag{2.48a}$$

$$= -k_{\mathrm{B}}T \ln \pi(\boldsymbol{x}) - k_{\mathrm{B}}T \ln Z \,, \tag{2.48b}$$

where $\delta\left[\boldsymbol{X}(\boldsymbol{\phi}) - \boldsymbol{x}\right]$ selects all microstates $\boldsymbol{\phi}$ with collective variable at point $\boldsymbol{x}$ and $Z = \sum_{\boldsymbol{\phi}} e^{-\beta E_{\mathrm{int}}(\boldsymbol{\phi})}$ is the partition function of the (unperturbed) equilibrium distribution. If the collective variable is a one-dimensional reaction coordinate (such as the committor), Eq. (2.48) is the free energy of the reaction shown in

Fig. 1.1, which incorporates both energetic and entropic aspects of the (many) system configurations consistent with the given value of the reaction coordinate.

Calculation of free energy in collective-variable space can be a computationally expensive task since sufficient sampling of the entire collective-variable space is needed to ensure converged estimates of free energy. For systems with metastable conformations, a trajectory initiated in one mesostate spends a (potentially) long time sampling microstates in that mesostate before overcoming the energy barrier to sample the other mesostate, and due to the timescale separation between the simulation timestep and reaction transitions, it becomes impractical to compute free energies via a brute-force approach.

Instead, enhanced-sampling methods use control parameters to more efficiently sample relevant regions of conformational space and compute free energies or other equilibrium properties [32]. Equilibrium samples are drawn independently from several control-parameter values that span the relevant collective-variable space. Because the control-parameter perturbations (and resulting equilibrium distributions (2.45)) that have been applied to the system are known, it is possible to reweight equilibrium samples taken from a range of control-parameter values to infer the distribution of the unperturbed system in collective-variable space [32, 33]. An analogous result was derived for nonequilibrium control protocols [34] where the system is driven along a single collective variable; the free energy along this driving protocol can be recovered by reweighting observations at each value of the collective variable using the work on the system up to that point in the protocol.

2.3.2 *Thermodynamics of Control Protocols*

Control protocols change control parameters $\boldsymbol{\lambda}$ between initial $\boldsymbol{\lambda}_{\text{initial}}$ and final $\boldsymbol{\lambda}_{\text{final}}$ values in finite time, so that entire trajectories are relevant rather than equilibrium samples from independent simulations performed at each control-parameter value.

The combination of microstate $\boldsymbol{\phi}$ and control parameter $\boldsymbol{\lambda}$ has an associated energy $V(\boldsymbol{\phi}, \boldsymbol{\lambda})$. The first law of thermodynamics requires that energy be conserved; therefore, changes in the system's energy must be accounted for as heat or work,

$$\mathrm{d}V(\boldsymbol{\phi}, \boldsymbol{\lambda}, t) = \sum_{\boldsymbol{\phi}} \left[\frac{\partial V(\boldsymbol{\phi}, \boldsymbol{\lambda}, t)}{\partial \boldsymbol{\lambda}} \frac{\partial \boldsymbol{\lambda}}{\partial t} + \frac{\partial V(\boldsymbol{\phi}, \boldsymbol{\lambda}, t)}{\partial \boldsymbol{\phi}} \frac{\partial \boldsymbol{\phi}}{\partial t} \right] \tag{2.49a}$$

$$= đ\mathcal{W}(\boldsymbol{\phi}, \boldsymbol{\lambda}, t) + đQ(\boldsymbol{\phi}, \boldsymbol{\lambda}, t) , \tag{2.49b}$$

where d denotes an exact differential and $đ$ an inexact differential. The work on the system is alternatively expressed as

$$đ\mathcal{W}(t) = -\sum_i đ\lambda_i(t) X_i(\boldsymbol{\phi}, \boldsymbol{\lambda}, t) \tag{2.50}$$

in terms of forces $X_i(\boldsymbol{\phi}, \boldsymbol{\lambda}, t) \equiv -\partial V(\boldsymbol{\phi}, \lambda, t)/\partial\lambda_i$ conjugate to respective control parameters λ_i.

The mean work during a protocol of duration Δt is

$$\langle \mathcal{W} \rangle_{\boldsymbol{\Lambda}} = -\int_0^{\Delta t} \mathrm{d}t \sum_i \dot{\lambda}_i(t) \langle X_i(\boldsymbol{\phi}, \boldsymbol{\lambda}, t) \rangle_{\boldsymbol{\Lambda}} , \tag{2.51}$$

where $\dot{\lambda}_i \equiv \partial\lambda_i/\partial t$ is the velocity of control parameter i, and $\langle \ldots \rangle_{\boldsymbol{\Lambda}}$ denotes a nonequilibrium average over the protocol, which depends on current control parameters and the entire history of control-parameter changes and system response.

When this process is done reversibly (quasistatically) so that the system is in equilibrium throughout the protocol, the mean conjugate force is related to changes in the equilibrium free energy, [35]

$$\langle X_i(\boldsymbol{\phi}, \boldsymbol{\lambda}, t) \rangle_{\boldsymbol{\Lambda}}^{\mathrm{rev}} = \langle X_i \rangle_{\boldsymbol{\lambda}(t)}^{\mathrm{eq}} = -\frac{\partial F^{\mathrm{eq}}(\boldsymbol{\lambda})}{\partial \lambda_i} , \tag{2.52}$$

and the work on the system is equal to the free energy difference between control-parameter endpoints (2.46):

$$\mathcal{W} = \int_0^{\Delta t} \mathrm{d}t \sum_i \dot{\lambda}_i(t) \frac{\partial F^{\mathrm{eq}}(\boldsymbol{\lambda})}{\partial \lambda_i} \tag{2.53a}$$

$$= \int_0^{\Delta t} \mathrm{d}t \frac{\mathrm{d}F^{\mathrm{eq}}(\boldsymbol{\lambda})}{\mathrm{d}t} \tag{2.53b}$$

$$= F^{\mathrm{eq}}(\boldsymbol{\lambda}_{\mathrm{final}}) - F^{\mathrm{eq}}(\boldsymbol{\lambda}_{\mathrm{initial}}) . \tag{2.53c}$$

This equality between work and free energy is true for any quasistatic protocol, regardless of the exact path taken through control-parameter space.

When the control parameters are tuned in finite duration Δt, on average excess work is done on the system above the equilibrium free energy change,

$$\langle \mathcal{W}^{\mathrm{ex}} \rangle_{\boldsymbol{\Lambda}} = -\int_0^{\Delta t} \mathrm{d}t \sum_i \dot{\lambda}_i(t) \langle \delta X_i(\boldsymbol{\phi}, \boldsymbol{\lambda}, t) \rangle \tag{2.54}$$

where $\delta X_i(\boldsymbol{\phi}, \boldsymbol{\lambda}, t)) \equiv X_i(\boldsymbol{\phi}, \boldsymbol{\lambda}, t) - \langle X_i \rangle_{\boldsymbol{\lambda}(t)}^{\mathrm{eq}}$ is defined as the deviation of instantaneous conjugate force from its equilibrium mean at control parameter $\boldsymbol{\lambda}$.

The excess work during a protocol has special significance for nonequilibrium driving protocols. Assuming the system begins in equilibrium at $\boldsymbol{\lambda}_{\mathrm{initial}}$ and relaxes to equilibrium at $\boldsymbol{\lambda}_{\mathrm{final}}$ after the end of the protocol, excess work is equal to the irreversible entropy production of the nonequilibrium driving protocol, quantifying the energy wasted during the process.

$$\beta\langle\mathcal{W}^{\text{ex}}\rangle_\Lambda \equiv \beta(\langle\mathcal{W}\rangle_\Lambda - \Delta F^{\text{eq}}) \tag{2.55a}$$

$$= \beta\langle\Delta V\rangle_\Lambda - \beta\langle Q\rangle_\Lambda - \beta\langle\Delta V\rangle^{\text{eq}} + \Delta H^{\text{sys,eq}} \tag{2.55b}$$

$$= \beta\delta\langle V(\Delta t)\rangle_\Lambda + \Delta H_\Lambda^{\text{env}} + \Delta H^{\text{sys,eq}}\,. \tag{2.55c}$$

In (2.55b), the first law is used to equate work with change in energy and heat flow, and the free energy difference is expressed as changes in energy and entropy. Equation (2.55c) equates heat with the environmental entropy change and defines the deviation of energy at the end of the protocol from its equilibrium value as $\delta\langle V(\Delta t)\rangle_\Lambda \equiv \langle V(\Delta t)\rangle_\Lambda - \langle V\rangle^{\text{eq}}_{\boldsymbol{\lambda}_{\text{final}}}$. When the system relaxes to equilibrium after the end of the protocol, $\delta\langle V(t \to \infty)\rangle_\Lambda = 0$. Also recognizing that $\Delta H_\Lambda^{\text{env}} + \Delta H^{\text{sys}} = \Delta H_\Lambda^{\text{irr}}$, the excess work is equal to irreversible entropy production,

$$\beta\langle\mathcal{W}^{\text{ex}}\rangle_\Lambda = \Delta H_\Lambda^{\text{irr}}\,. \tag{2.56}$$

The connection between excess work and entropy production plays a central role in *nonequilibrium work relations*, such as the Jarzynski equality Eq. (1.9) [36] which allows estimation of the equilibrium free-energy difference ΔF^{eq} from nonequilibrium driving experiments, and Crooks' work fluctuation theorem Eq. (1.10) [37] which places a tight constraint on allowable work fluctuations.

2.3.3 Linear-Response Approximation to Excess Work

Given the role excess work plays in the nonequilibrium thermodynamics of driving protocols, it is helpful to understand how it depends on the choice of driving protocol. To this end, results have been derived to exactly or approximately calculate the excess work for different types of protocols. Exact results are analytically tractable for only a few special cases, such as for driving a system in a harmonic trap with control over the trap center and spring constant [29]. Since exact results are difficult to derive, it is valuable to make approximations to the protocol work valid in different limits. In this thesis, the long-duration limit of protocol work is examined, using an approximation of excess work that uses linear-response theory.

The mean excess power on a system at time t is

$$\langle\mathcal{P}^{\text{ex}}(t)\rangle_{\boldsymbol{\Lambda}} = -\sum_i \dot{\lambda}_i(t)\langle\delta X_i(t)\rangle^{\text{neq}}\,, \tag{2.57}$$

the product of the (nonequilibrium) value of the conjugate forces and the rate of change of each control parameter at a given time. For a given control protocol, $\dot{\boldsymbol{\lambda}}(t)$ is fixed and the nonequilibrium lag of conjugate forces from their equilibrium values throughout the control protocol must be determined. In Ref. [38], the lag is approximated using linear-response theory, which is outlined below.

For quasistatic protocols, the conjugate forces relax to their quasistatic values throughout the protocol. Assuming the protocols are performed in long-but-finite duration and that control parameters change continuously with smooth first and second derivatives, the system response to external perturbations can be approximated using linear-response theory [2] as

$$\langle \delta X_i(t) \rangle^{\text{neq}} \approx \int_{-\infty}^{t} \mathrm{d}t' \sum_j \chi_{ij}^{\boldsymbol{\lambda}(t)}(t-t') \left[\lambda_j(t') - \lambda_j(t) \right] . \tag{2.58}$$

Here, $\chi_{ij}^{\boldsymbol{\lambda}}(t-t')$ is the *linear-response function*

$$\chi_{ij}^{\boldsymbol{\lambda}}(t-t') \equiv \beta \frac{\mathrm{d}}{\mathrm{d}t} \langle \delta X_j(t') \delta X_i(t) \rangle_{\boldsymbol{\lambda}} \tag{2.59}$$

representing the response of conjugate force X_i at time t to a step perturbation of control parameter λ_j at time t' at control-parameter value $\boldsymbol{\lambda}$.

Integration by parts yields an approximation for the nonequilibrium conjugate force as

$$\langle \delta X_i(t) \rangle^{\text{neq}} \approx \int_{-\infty}^{t} \mathrm{d}t' \sum_j \langle \delta X_i(0) \delta X_j(t-t') \rangle \dot{\lambda}_j(t') . \tag{2.60}$$

Assuming the control-parameter velocity changes slowly, a Taylor expansion of the control-parameter velocity $\dot{\boldsymbol{\lambda}}(t')$ around $t' = t$, truncated at the constant term yields

$$\langle \delta X_i(t) \rangle^{\text{neq}} \approx -\sum_j \zeta_{ij}(\boldsymbol{\lambda}(t)) \dot{\lambda}_j(t) , \tag{2.61}$$

where

$$\zeta_{ij}(\boldsymbol{\lambda}) \equiv \int_0^{\infty} \mathrm{d}t \, \langle \delta X_i(0) \delta X_j(t) \rangle_{\boldsymbol{\lambda}} \tag{2.62}$$

is the *generalized friction matrix*, the time integral of the correlation function of conjugate forces for the system *at equilibrium* at control parameter $\boldsymbol{\lambda}$.

Combining this approximation for the mean nonequilibrium lag of conjugate forces with the control-parameter velocity yields the linear-response estimate for the mean excess power throughout the protocol, which is valid in the long-duration limit,

$$\mathcal{P}^{\text{ex,LR}}(t) = \sum_{ij} \dot{\lambda}_i(t) \zeta_{ij}(\boldsymbol{\lambda}(t)) \dot{\lambda}_j(t) . \tag{2.63}$$

This phenomenological expression for the excess power is local, depending only on instantaneous values of the control parameters and their time derivatives, rather than depending on the entire history of control-parameter perturbations. Integrating excess power over the protocol duration yields the mean excess work

$$\langle \mathcal{W}^{\mathrm{ex}} \rangle_{\boldsymbol{\Lambda}} \approx \int_0^{\Delta t} \mathrm{d}t \, \mathcal{P}^{\mathrm{ex,LR}}(t) \, . \tag{2.64}$$

2.3.4 Theoretical Insight from Generalized Friction Metric

The linear-response approximation to the excess work during a protocol has several features that provide insight into minimum-work protocols in the long-duration limit.

First, the generalized friction matrix has special mathematical properties that make it a Riemannian metric, providing a thermodynamic measure of the distance between two equilibrium distributions in control-parameter space which accounts for fluctuations required for the system to relax between the two distributions [35]. Minimum-work protocols are *geodesics* of the generalized friction in control-parameter space, which are shortest paths when distance is measured in terms of the friction metric rather than Euclidean projections. The friction defines a thermodynamic length for a given protocol,

$$\mathcal{L}_{\boldsymbol{\Lambda}} = \int_0^{\Delta t} \mathrm{d}t \sqrt{\mathcal{P}^{\mathrm{ex,LR}}(t)} \, , \tag{2.65}$$

which is independent of duration Δt. Using the Cauchy-Schwarz inequality, thermodynamic length provides a lower bound on protocol excess work

$$\mathcal{W}_{\boldsymbol{\Lambda}}^{\mathrm{ex,LR}} = \int_0^{\Delta t} \mathrm{d}t \, \mathcal{P}^{\mathrm{ex,LR}}(t) \tag{2.66a}$$

$$= \frac{1}{\Delta t} \int_0^{\Delta t} \mathrm{d}t \, 1 \cdot \int_0^{\Delta t} \mathrm{d}t \left[\sqrt{\mathcal{P}^{\mathrm{ex,LR}}(t)} \right]^2 \tag{2.66b}$$

$$\geq \frac{1}{\Delta t} \left[\int_0^{\Delta t} \mathrm{d}t \sqrt{\mathcal{P}^{\mathrm{ex,LR}}(t)} \right]^2 \tag{2.66c}$$

$$= \frac{\mathcal{L}^2}{\Delta t} \, . \tag{2.66d}$$

This lower bound is realized for protocols with constant excess power [38]. Therefore, minimum-work protocols in the linear-response regime are independent of protocol duration (up to a constant scaling of control-parameter velocity) and have constant excess power.

The generalized friction can be decomposed into components measuring the size of force fluctuations and the timescale of relaxation in the equilibrium distribution at control parameter $\boldsymbol{\lambda}$:

$$\zeta_{ij}(\boldsymbol{\lambda}) = \langle \delta X_i \delta X_j \rangle_{\boldsymbol{\lambda}} \tau_{ij}(\boldsymbol{\lambda}) \,, \tag{2.67}$$

where $\langle \delta X_i \delta X_j \rangle_{\boldsymbol{\lambda}}$ is the equilibrium covariance between conjugate force X_i and X_j at control parameter $\boldsymbol{\lambda}$, and

$$\tau_{ij}(\boldsymbol{\lambda}) = \int_0^{\infty} \mathrm{d}t \, \frac{\langle \delta X_i(0) \delta X_j(t) \rangle_{\boldsymbol{\lambda}}}{\langle \delta X_i \delta X_j \rangle_{\boldsymbol{\lambda}}} \tag{2.68}$$

is the integral relaxation time between conjugate forces X_i and X_j. The relevance of the force covariance matrix to control protocols is discussed in Ref. [35].

A major advantage of the linear-response approximation to excess work is the local description of the conjugate force lag from local perturbations. This greatly simplifies the original situation where the nonequilibrium conjugate force depends on the entire protocol history. The protocol that minimizes the linear-response approximation of excess work can be found using calculus of variations [1]. This amounts to solving the Euler-Lagrange equation [1] for $\boldsymbol{\lambda}(t)$

$$\zeta_{ki} \ddot{\lambda}_i + \frac{\partial \zeta_{ki}}{\partial \lambda_j} \dot{\lambda}_j \dot{\lambda}_i - \frac{1}{2} \frac{\partial \zeta_{ij}}{\partial \lambda_k} \dot{\lambda}_j \dot{\lambda}_i = 0. \tag{2.69}$$

with endpoints $\boldsymbol{\lambda}(0) = \boldsymbol{\lambda}_{\text{initial}}$ and $\boldsymbol{\lambda}(\Delta t) = \boldsymbol{\lambda}_{\text{final}}$. For most systems, this Euler-Lagrange equation is not exactly solvable. Section 2.3.5 provides a method for generating numerical solutions to this equation. However, for a one-dimensional control parameter λ, the Euler-Lagrange equation can be solved for arbitrary generalized friction [38], yielding

$$\dot{\lambda}^{\text{MWP}}(t) \propto \zeta(\lambda(t))^{-1/2} \tag{2.70}$$

where the proportionality constant is chosen such that the protocol is completed in duration Δt. The resulting design principle of one-dimensional minimum-work control says that minimizing work requires slower changes to the control parameter wherever the resistance to driving (i.e., generalized friction) is large. The performance of one-dimensional driving using this framework have been studied numerically in several model systems [39–41].

2.3.5 *Numerical Calculations of Minimum-Work Protocols*

Solving the Euler-Lagrange equation for minimum-work protocols (Eq. (2.69)) is a non-trivial task. Analytical forms for the generalized friction are only available for a handful of systems [38, 39, 42], necessitating numerical estimation of the generalized friction at a range of control-parameter values. Optimization of a one-dimensional protocol is fairly straightforward since the path is fixed and only the control-parameter velocity must be optimized. This can be done by numerically estimating the friction coefficient for a discrete set of M control parameter values $\lambda(m)$ between protocol endpoints, then solving Eq. (2.70) to find the time $t^{\mathrm{MWP}}(\lambda(m))$ for each control-parameter value $\lambda(m)$ during the protocol. Beginning with $t^{\mathrm{MWP}}(\lambda_0) = 0$, subsequent times can be found by expressing Eq. (2.70) as an iterative relation of discrete time intervals

$$t^{\mathrm{MWP}}(\lambda(m+1)) = t^{\mathrm{MWP}}(\lambda(m)) + \mathrm{d}t^{\mathrm{MWP}}_{m\to m+1} \tag{2.71a}$$

$$= t^{\mathrm{MWP}}(\lambda(m)) + \left[\frac{\zeta(\lambda(m)) + \zeta(\lambda(m+1))}{2}\right]^{1/2} [\lambda(m+1) - \lambda(m)] \ . \tag{2.71b}$$

All times $t^{\mathrm{MWP}}(\lambda(m))$ can be divided by the final time $t^{\mathrm{MWP}}(\lambda(M-1))$ to obtain the optimal scaled protocol duration for each control-parameter point. This algorithm gives the protocol time for control-parameter points with equal spatial separation; in Appendix B.2.1, a method to calculate a one-dimensional minimum-work protocol with discrete points equally spaced in time is discussed.

Designing multidimensional control protocols is more involved, since the spatial path through multidimensional control-parameter space must also be optimized. Additionally, it is typically impractical to sample the friction matrix over control-parameter space for more than a few control parameters. A few numerical methods have been developed to calculate minimum-work protocols, some relying on the generalized friction framework [43, 44] and others on importance sampling [45] or machine-learning approaches [46]. An approach based on the string method found in Ref. [44] is outlined below. The string method was originally developed to find minimum-energy paths between metastable states in configuration space [24].

The string method divides a control protocol into M discrete segments, with each node at a point in control-parameter space. The string is initialized as a straight line between control-parameter endpoints, then iteratively updated to solve a discrete version of the Euler-Lagrange equation (2.69). The string is updated from the nth to $(n+1)$th iteration by solving the set of linear equations for each component λ_α at each discrete point m along the string:

$$\lambda_\alpha^{n+1}(m) - \lambda_\alpha^n(m) = \Delta r \Bigg\{ \mathrm{D}^2\lambda_\alpha^{n+1}(m) + \sum_{ijk} [\zeta^{-1}]^n_{\alpha k}(\boldsymbol{\lambda}(m))\, \mathrm{D}\lambda_i^n(m)\, \mathrm{D}\lambda_j^n(m) \left[\partial_i \zeta_{kj}(\boldsymbol{\lambda}(m)) - \frac{1}{2}\partial_k \zeta_{ij}(\boldsymbol{\lambda}(m))\right]\Bigg\} \tag{2.72}$$

where $\partial_i \zeta_{kj}(\boldsymbol{\lambda}) = \partial \zeta_{kj}(\boldsymbol{\lambda})/\partial \lambda_i$ is the partial derivative of the kjth component of the friction matrix with respect to λ_i (assuming all other control parameters are fixed) at control-parameter value $\boldsymbol{\lambda}$, Δr is a parameter controlling the size of the string update, and

$$\mathrm{D}\lambda_\alpha(m) \equiv \frac{\lambda_\alpha(m+1) - \lambda_\alpha(m-1)}{2\delta t} \tag{2.73a}$$

$$\mathrm{D}^2\lambda_\alpha(m) \equiv \frac{\lambda_\alpha(m+1) + \lambda_\alpha(m-1) - 2\lambda_\alpha(m)}{\delta t^2} \tag{2.73b}$$

are finite-difference estimators of time derivatives for the string, with δt denoting the difference in scaled protocol time between adjacent string points. The string is repeatedly updated using Eq. (2.72) until it converges to user-specified precision.

The string update step requires estimation of the friction matrix and its derivative in each control-parameter direction for each point in the current iteration of the string. If these quantities can be estimated from short equilibrium simulations at each control-parameter value, the estimates can be obtained in parallel to reduce computational cost. The estimates from all string points are inserted into Eq. (2.72) to update the string. However, accurate calculation of the friction matrix and its derivative for a single control-parameter value can take significant sampling time. If the friction matrix has constant integral relaxation time for all components across control-parameter space, the force variance matrix $\zeta_{ij}(\boldsymbol{\lambda}) \approx \langle \delta X_j \delta X_i \rangle_{\boldsymbol{\lambda}}$ takes the place in Eq. (2.64) of the friction matrix [44]. In this case, the derivative of the force variance matrix in control-parameter space is $\partial_i \langle \delta X_j \delta X_k \rangle_{\boldsymbol{\lambda}} = \langle \delta X_i \delta X_j \delta X_k \rangle_{\boldsymbol{\lambda}}$. Estimating second and third cumulants from an equilibrium distribution take significantly less sampling time than estimating the integrated correlation function of the conjugate forces; if the constant-relaxation-time approximation holds, there is a significant reduction in computational cost for computing minimum-work protocols using this method. This approximation was used in [44] to design minimum-work protocols in a 100-dimensional control-parameter space.

Minimum-work protocols are calculated using the string method in Chap. 4, with some modifications to account for the varying integral relaxation time throughout control-parameter space for the chosen model system, that precludes using the force variance matrix alone to design minimum-work protocols.

References

1. Morin, D.: Introduction to Classical Mechanics. Cambridge University Press, Cambridge (2008)
2. Zwanzig, R.: Nonequilibrium Statistical Mechanics. Oxford University Press, New York (2001)
3. Peliti, L., Pigolotti, S.: Stochastic Thermodynamics: An Introduction. Princeton University Press, Princeton (2021)

4. Frenkel, D., Smit, B.: Understanding Molecular Simulation: From Algorithms to Applications, 2nd edn. Academic Press, San Diego (2002)
5. Press, W., Teukolksy, S.A., Vetterline, W.T., Flannery, B.P.: Numerical Recipes: The Art of Scientific Computing. Cambridge University Press, Cambridge (2007)
6. Glauber, R.J.: Time-dependent statistics of the Ising model. J. Math. Phys. **4**(2), 294–307 (1963)
7. Cover, T.M., Thomas, J.A.: Elements of Information Theory, 2nd edn. Wiley, Hoboken (2006)
8. Landauer, R.: Irreversibility and heat generation in the computing process. IBM J. Res. Dev. **5**, 183–191 (1961)
9. Atkins, P., de Paula, J., Keeler, J.: Physical Chemistry, 11th edn. Oxford University Press, Oxford (2018)
10. Horowitz, J.M.: Multipartite information flow for multiple Maxwell demons. J. Stat. Mech. Theory Exp. **2015**(3), 1–13 (2015)
11. Hartich, D., Barato, A., Seifert, U.: Stochastic thermodynamics of bipartite systems: transfer entropy inequalities and a Maxwell's demon interpretation. J. Stat. Mech. Theory Exp. **2014**, 02016 (2014)
12. Barato, A.C., Hartich, D., Seifert, U.: Rate of mutual information between coarse-grained non-Markovian variables. J. Stat. Phys. **153**, 460–478 (2013)
13. Van Den Broeck, C., Esposito, M.: Ensemble and trajectory thermodynamics: a brief introduction. Physica A **418**, 6–16 (2015)
14. Brown, A.I., Sivak, D.A.: Theory of nonequilibrium free energy transduction by molecular machines. Chem. Rev. **120**(1), 434–459 (2020)
15. Seifert, U.: From stochastic thermodynamics to thermodynamic inference. Annu. Rev. Condens. Matter Phys. **10**, 171–192 (2019)
16. Metzner, P., Schutte, C., Vanden-Eijnden, E.: Transition path theory for Markov jump processes. Multiscale Model. Simul. **7**(3), 1192–1219 (2009)
17. Vanden-Eijnden, E.: Transition path theory. In: Bowman, G.R., Pande, V.S., Noe, F. (eds.) An Introduction to Markov State Models and Their Application to Long Timescale Molecular Simulation, chapter 7, pp. 91–100. Springer, Berlin (2014)
18. Weinan, E., Vanden-Eijnden, E.: Towards a theory of transition paths. J. Stat. Phys. **123**(3), 503–523 (2006)
19. Berezhkovskii, A.M., Szabo, A.: Committors, first-passage times, fluxes, Markov states, milestones, and all that. J. Chem. Phys **150**, 54106 (2019)
20. Peters, B.: Reaction coordinates and mechanistic hypothesis tests. Annu. Rev. Phys. Chem. **67**, 669–690 (2016)
21. Bolhuis, P.G., Dellago, C.: Practical and conceptual path sampling issues. Eur. Phys. J. Spec. Top. **224**, 2409–2427 (2015)
22. Berezhkovskii, A., Szabo, A.: One-dimensional reaction coordinates for diffusive activated rate processes in many dimensions. J. Chem. Phys. **122**, 014503 (2005)
23. Vanden-Eijnden, E., Venturoli, M.: Revisiting the finite temperature string method for the calculation of reaction tubes and free energies. J. Chem. Phys. **130**(19), 194103 (2009)
24. Weinan, E., Ren, W., Vanden-Eijnden, E.: String method for the study of rare events. Phys. Rev. B **66**, 052301 (2002)
25. Maragliano, L., Fischer, A., Vanden-Eijnden, E., Ciccotti, G.: String method in collective variables: minimum free energy paths and isocommittor surfaces. J. Chem. Phys. **125**(2), 024106 (2006)
26. Zhao, R., Shen, J., Skeel, R.D.: Maximum flux transition paths of conformational change. J. Chem. Theory Comput. **6**(8), 2411–2423 (2010)
27. Johnson, M.E., Hummer, G.: Characterization of a dynamic string method for the construction of transition pathways in molecular reactions. J. Phys. Chem. B **116**(29), 8573–8583 (2012)
28. Venturoli, M., Vanden-Eijnden, E., Ciccotti, G.: Kinetics of phase transitions in two dimensional Ising models studied with the string method. J. Math. Chem. **45**, 188–222 (2009)
29. Schmiedl, T., Seifert, U.: Optimal finite-time processes in stochastic thermodynamics. Phys. Rev. Lett. **98**(10), 1–4 (2007)

30. Sivak, D.A., Crooks, G.E.: Thermodynamic geometry of minimum-dissipation driven barrier crossing. Phys. Rev. E **94**, 052106 (2016)
31. Blaber, S., Sivak, D.A.: Skewed thermodynamic geometry and optimal free energy estimation. J. Chem. Phys. **153**(24), 244119 (2020)
32. Tiwary, P., van de Walle, A.: A review of enhanced sampling approaches for accelerated molecular dynamics. In: Weinberger, C.R., Tucker, G.J. (eds.) Multiscale materials modeling for nanomechanics, chapter 6, pp. 195–221. Springer International Publishing, Berlin (2016)
33. Kumar, S., Rosenberg, J.M., Bouzida, D., Swendsen, R.H., Kollman, P.A.: The weighted histogram analysis method for free-energy calculations on biomolecules. I. The method. J. Comput. Chem. **13**, 1011–1021 (1992)
34. Hummer, G., Szabo, A.: Free energy reconstruction from nonequilibrium single-molecule pulling experiments. Proc. Natl. Acad. Sci. U. S. A. **98**(7), 3658–3661 (2001)
35. Crooks, G.E.: Measuring thermodynamic length. Phys. Rev. Lett. **99**, 100602 (2007)
36. Jarzynski, C.: Nonequilibrium equality for free energy differences. Phys. Rev. Lett. **78**(14), 2690–2693 (1997)
37. Crooks, G.E.: Path-ensemble averages in systems driven far from equilibrium. Phys. Rev. E **61**(3), 2361–2366 (2000)
38. Sivak, D.A., Crooks, G.E.: Thermodynamic metrics and optimal paths. Phys. Rev. Lett. **108**(19), 190602 (2012)
39. Zulkowski, P.R., DeWeese, M.R.: Optimal control of overdamped systems. Phys. Rev. E **92**, 032117 (2015)
40. Lucero, J.N.E., Mehdizadeh, A., Sivak, D.A.: Optimal control of rotary motors. Phys. Rev. E **99**, 012119 (2019)
41. Blaber, S., Sivak, D.A.: Optimal control of protein copy number. Phys. Rev. E **101**, 22118 (2020)
42. Zulkowski, P.R., Sivak, D.A., Crooks, G.E., Deweese, M.R.: Geometry of thermodynamic control. Phys. Rev. E - Stat. Nonlinear Soft Matter Phys. **86**(4), 1–8 (2012)
43. Rotskoff, G.M., Crooks, G.E.: Optimal control in nonequilibrium systems: dynamic Riemannian geometry of the Ising model. Phys. Rev. E **92**, 060102 (2015)
44. Rotskoff, G.M., Crooks, G.E., Vanden-Eijnden, E.: A geometric approach to optimal nonequilibrium control: minimizing dissipation in nanomagnetic spin systems. Phys. Rev. E **95**, 012148 (2017)
45. Gingrich, T., Rotskoff, G., Crooks, G., Geissler, P.: Near-optimal protocols in complex nonequilibrium transformations. PNAS **113**, 10263–10268 (2016)
46. Engel, M.C., Smith, J.A., Brenner, M.P.: Optimal control of nonequilibrium systems through automatic differentiation (2022). arXiv:2201.00098

Chapter 3
Information Thermodynamics of Transition Paths

Abstract In this chapter, an alternative perspective of transition-path theory (which describes reactive events) is developed that unites it with stochastic thermodynamics to describe flows of entropy, energy, and information during the reaction. The main result is a quantification of the irreversibility of system dynamics while a reaction occurs, with parallel definitions of an entropy production rate and mutual information flow for transition-path dynamics. These provide a thermodynamic measure of the relevance of a particular degree of freedom to the reaction, yielding an optimization criterion for selecting collective variables that describe a reaction. (Material in this chapter also appears in Louwerse and Sivak (Phys Rev Lett 128:170602, 2022).)

The reaction coordinate describing a transition between reactant and product is a fundamental concept in the theory of chemical reactions. As discussed in Sect. 1.1.2, the committor is widely considered to be the "true" reaction coordinate, describing every aspect of the transition from $A \to B$. In this chapter, a novel information-theoretic justification of the committor as the reaction coordinate is derived. Additionally, it is shown that selecting the transition-path ensemble (the set of trajectories from reactant to product) from a long ergodic equilibrium trajectory results in entropy production that precisely equals the information generated by system dynamics about the reactivity of trajectories. The components of entropy production and information generation due to an arbitrary system coordinate are also equal; this reveals equivalent thermodynamic and information-theoretic measures of the suitability of low-dimensional collective variables that encode information relevant for describing reaction mechanisms, the need for which was discussed in Sect. 2.2. The committor is a single coordinate that preserves all system entropy production and distills all system information about reactivity, giving further support for its role as the reaction coordinate. The equality between entropy production and dynamical information generation also holds at the level of arbitrary individual coordinates, providing parallel measures of the coordinate's relevance to the reaction. The results derived in this chapter use ideas from the fields of information theory (Sect. 2.1.2), stochastic thermodynamics (Sect. 2.1.3), and transition-path theory (Sect. 2.2).

M. Louwerse, *Efficient Control and Spontaneous Transitions*, Springer Theses,
https://doi.org/10.1007/978-3-031-40534-1_3

3.1 Information-Theoretic Perspective on the Transition Path Ensemble

Consider a multidimensional system $\boldsymbol{\Phi}$ evolving according to Markovian dynamics governed by the Master equation [1], $\mathrm{d}_t p(\boldsymbol{\phi}) = \sum_{\boldsymbol{\phi}'} T_{\boldsymbol{\phi}\boldsymbol{\phi}'} p(\boldsymbol{\phi}')$, where $T_{\boldsymbol{\phi}\boldsymbol{\phi}'}$ is the transition rate from state $\boldsymbol{\phi}' \to \boldsymbol{\phi}$ and $p(\boldsymbol{\phi})$ is the probability of state $\boldsymbol{\phi}$. It is assumed that the transition rates obey detailed balance [1] and the system is in equilibrium with its environment so that $p(\boldsymbol{\phi}) = \pi(\boldsymbol{\phi})$, the equilibrium probability of $\boldsymbol{\phi}$. The transition-path ensemble (TPE) is studied, the set of trajectories that leave one subset of states $A \in \boldsymbol{\Phi}$ and next visit a distinct subset $B \in \boldsymbol{\Phi} \setminus A$ before returning to A. In most applications, A and B are metastable states separated by a dynamical barrier; following Refs. [2, 3] it is only assumed that A and B do not overlap and lack direct transitions; i.e., $T_{\boldsymbol{\phi}\boldsymbol{\phi}'} = 0$ for $\boldsymbol{\phi}' \in A$ and $\boldsymbol{\phi} \in B$.

The TPE is formed by selecting from a long ergodic equilibrium *supertrajectory* the trajectory segments that leave A and reach B before A. Transition paths are therefore selected based on the trajectory *outcome* S_+ [the next mesostate (A or B) visited by the system] and *origin* S_- (the mesostate most recently visited by the system). This partitions the supertrajectory into four trajectory subensembles, each with particular $\boldsymbol{s} \equiv (s_-, s_+)$: The forward (reverse) transition-path ensemble is the set of trajectory segments with $\boldsymbol{s} = (A, B)$ [$\boldsymbol{s} = (B, A)$], and the stationary subensemble from $A \to A$ ($B \to B$) has $\boldsymbol{s} = (A, A)$ [$\boldsymbol{s} = (B, B)$], as depicted in Fig. 3.1. Every trajectory segment in the forward TPE has a corresponding equally probable time-reversed trajectory segment in the reverse TPE.

At any time during the equilibrium supertrajectory, random variables $\boldsymbol{\Phi}$ and $\boldsymbol{S}$ are defined that respectively denote the current system state and trajectory subensemble, with $p(\boldsymbol{\phi}, \boldsymbol{s})$ the joint distribution that the system is currently in state $\boldsymbol{\phi}$ and is currently on a trajectory segment with respective origin and outcome $\boldsymbol{s} = \{s_-, s_+\}$. Since the system dynamics are Markovian, the trajectory outcome and origin are conditionally independent given current state $\boldsymbol{\phi}$, so the joint distribution can be factored as [2]

$$p(\boldsymbol{\phi}, \boldsymbol{s}) = \pi(\boldsymbol{\phi}) p(s_+|\boldsymbol{\phi}) p(s_-|\boldsymbol{\phi}) . \tag{3.1}$$

The conditional probabilities of trajectory outcome and origin given current state $\boldsymbol{\phi}$ are

$$p(S_+ = B \mid \boldsymbol{\phi}) = q^+(\boldsymbol{\phi}) \tag{3.2a}$$

$$p(S_- = A \mid \boldsymbol{\phi}) = q^-(\boldsymbol{\phi}) . \tag{3.2b}$$

Here $q^+(\boldsymbol{\phi})$ is the forward *committor*, the probability that the system currently in state $\boldsymbol{\phi}$ will next reach B before A, and $q^-(\boldsymbol{\phi})$ is the backward committor, the probability that the system (currently in $\boldsymbol{\phi}$) was more recently in mesostate A than in B. The committors obey boundary conditions $q^+(\boldsymbol{\phi}) = 0$ and $q^-(\boldsymbol{\phi}) = 1$ for $\boldsymbol{\phi} \in A$,

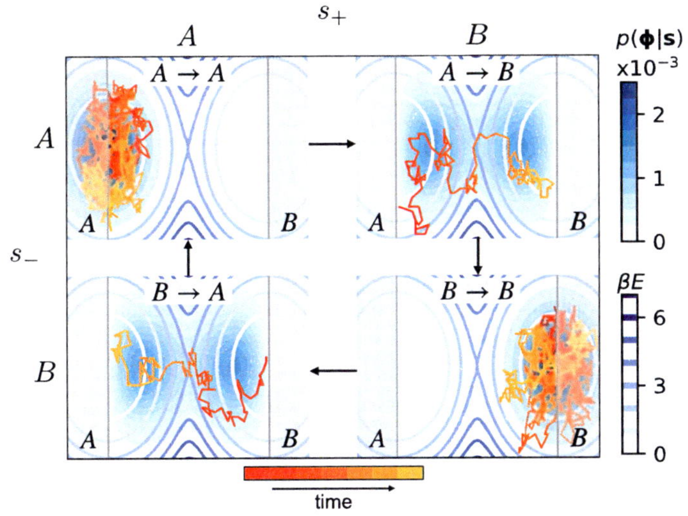

Fig. 3.1 Partitioning a long ergodic equilibrium supertrajectory into subensembles based on trajectory outcome S_+ and origin S_-. Contours: example double-well potential energy βE_{int}. Shaded regions: A is the region to the left of the left energy minimum and B is the region to the right of the right minimum. Heat map: probability distribution $p(\boldsymbol{\phi}|s)$ of system state conditioned on trajectory subensemble s. Solid curves: representative trajectory from each subensemble. The forward (reverse) TPE in the top-right (bottom-left) panel has net flux of trajectories from $A \to B$ ($B \to A$). The top-left (bottom-right) panel shows the stationary subensemble from $A \to A$ ($B \to B$)

and $q^+(\boldsymbol{\phi}) = 1$ and $q^-(\boldsymbol{\phi}) = 0$ for $\boldsymbol{\phi} \in B$. Since the system is in equilibrium and the transition rates obey detailed balance, $q^-(\boldsymbol{\phi}) = 1 - q^+(\boldsymbol{\phi})$ [2], a single committor (without loss of generality, the forward committor $q^+(\boldsymbol{\phi})$) provides information about both the outcome and origin of the trajectory segment, so $q^+(\boldsymbol{\phi})$ is referred to as the reaction coordinate.

During the equilibrium supertrajectory, the system continually evolves from A to B and B to A, completing a unidirectional cycle through each trajectory subensemble with stochastic transition times depending on underlying microscopic dynamics. Transition-path theory [3–5] derives quantitative properties (reaction rate and free-energy difference) of the $A \to B$ reaction from the equilibrium probability flux of subensemble transitions

$$\nu_S = \sum_{\boldsymbol{\phi} \notin A, \boldsymbol{\phi}' \in A} T_{\boldsymbol{\phi}\boldsymbol{\phi}'} \pi(\boldsymbol{\phi}') q^+(\boldsymbol{\phi}) \tag{3.3}$$

and the respective marginal probabilities $p(s_+)$ and $p(s_-)$:

$$k_{AB} = \frac{\nu_S}{p(S_- = A)} = \frac{\nu_S}{p(S_+ = A)} \tag{3.4a}$$

$$k_{BA} = \frac{\nu_S}{p(S_- = B)} = \frac{\nu_S}{p(S_+ = B)} \tag{3.4b}$$

$$\beta \Delta F_{AB} = \ln \frac{p(S_- = A)}{p(S_- = B)} = \ln \frac{p(S_+ = A)}{p(S_+ = B)}, \tag{3.4c}$$

where k_{AB} (k_{BA}) is the rate constant for the $A \to B$ ($B \to A$) transition and $\Delta F_{AB} \equiv F_B - F_A$ is the free-energy difference. The second equality in each line uses $p(S_- = A) = p(S_+ = A)$ which follows from detailed balance. Mesoscopic reaction properties are thus derived from information about the subensembles, specifically the proportion of time spent in each subensemble and how frequently the subensemble switches.

Therefore, the reaction coordinate should be maximally informative about the current subensemble. This is precisely quantified by mutual information, a non-linear statistical measure of the relationship between two random variables, specifically quantifying the reduction of uncertainty (given by Shannon entropy $H(X) \equiv -\sum_x p(x) \ln p(x)$) about one random variable from measuring another [6]:

$$I(S; \mathbf{\Phi}) \equiv \sum_{\boldsymbol{\phi}, s} p(\boldsymbol{\phi}, s) \ln \frac{p(\boldsymbol{\phi}, s)}{\pi(\boldsymbol{\phi}) p(s)}, \tag{3.5}$$

where $p(s) = \sum_{\boldsymbol{\phi}} p(\boldsymbol{\phi}, s)$ is the marginal probability that the system is currently on a trajectory segment with outcome and origin $s = (s_-, s_+)$. Operationally, $p(s)$ can be estimated from the proportion of time τ_s spent in subensemble s during an equilibrium supertrajectory of length τ, $p(s) = \lim_{\tau \to \infty} \tau_s / \tau$. The mutual information between trajectory outcome and origin and a general state function $f(\boldsymbol{\phi})$ is constrained by the data processing inequality [6] such that $I(S; \mathbf{\Phi}) \geq I(S; f(\mathbf{\Phi}))$. The committor satisfies this equality:

$$I(S; \mathbf{\Phi}) = \sum_{\boldsymbol{\phi}, s} \pi(\boldsymbol{\phi}) p(s|\boldsymbol{\phi}) \ln \frac{p(s|\boldsymbol{\phi})}{p(s)} \tag{3.6a}$$

$$= \sum_{\boldsymbol{\phi}, s} \sum_{q_o} \pi(\boldsymbol{\phi}) p(s|\boldsymbol{\phi}) \ln \frac{p(s|\boldsymbol{\phi})}{p(s)} \delta[q^+(\boldsymbol{\phi}) - q_o] \tag{3.6b}$$

$$= \sum_{q_o, s} p(s|q_o) \ln \frac{p(s|q_o)}{p(s)} \left[\sum_{\boldsymbol{\phi}} \pi(\boldsymbol{\phi}) \delta[q^+(\boldsymbol{\phi}) - q_o] \right] \tag{3.6c}$$

$$= \sum_{q_o, s} \pi(q_o) p(s|q_o) \ln \frac{p(s|q_o)}{p(s)} \tag{3.6d}$$

$$= \sum_{q_o, s} p(q_o, s) \ln \frac{p(q_o, s)}{\pi(q_o) p(s)} \tag{3.6e}$$

$$= I(S; q^+(\boldsymbol{\Phi})) \,. \tag{3.6f}$$

In (3.6a), the mutual information is expressed in terms of conditional probability $p(s|\boldsymbol{\phi})$. In (3.6b), a sum over all discrete committor values appearing in the system is included, allowing separation of all states with the same committor value in Eq. (3.6c), where it is also recognized that the probability the system is in a particular subensemble given state $\boldsymbol{\phi}$ depends only on its committor value, $p(s|\boldsymbol{\phi}) = p(s|q^+(\boldsymbol{\phi}))$. This allows the mutual information between trajectory subensemble and system state to be expressed as the mutual information between trajectory subensemble and the committor of system state in Eq. (3.6f). Since $I(S; \boldsymbol{\Phi}) = I(S; q^+(\boldsymbol{\Phi}))$, the committor is a sufficient statistic for the mutual information between trajectory subensemble and full system state [6]. In this sense, the committor is the "optimal" reaction coordinate, since it is maximally informative about the trajectory subensemble given a measurement of system state. This is the first major result.

3.1.1 Physical Origin of Subensemble Random Variables

Physically, the trajectory outcome and origin (and hence the committors) represent uncertainty in the state of the environment. Classical mechanics assumes a constant-energy universe (system $\boldsymbol{\Phi}$ plus environment $\boldsymbol{\Psi}$) governed by deterministic dynamics so that the outcome and origin of the trajectory initiated from a given state of system *and* environment are deterministic (and can be determined by integrating the state of the universe forward and backward in time until the system reaches A or B), i.e., $p(s|\boldsymbol{\phi}, \boldsymbol{\psi})$ is either 0 or 1. This partitions the state space of the universe into four quadrants corresponding to each trajectory subensemble, with each state $(\boldsymbol{\phi}, \boldsymbol{\psi})$ belonging to only one subensemble; thus the uncertainty about the trajectory subensemble given a state of the universe is zero,

$$H(S|\boldsymbol{\Phi}, \boldsymbol{\Psi}) \equiv -\sum_{\boldsymbol{\phi}, \boldsymbol{\psi}, s} p(\boldsymbol{\phi}, \boldsymbol{\psi}, s) \ln p(s|\boldsymbol{\phi}, \boldsymbol{\psi}) = 0. \tag{3.7}$$

In this case, the mutual information between the universe and trajectory subensemble is the uncertainty about the trajectory subensemble,

$$I(S; \boldsymbol{\Phi}, \boldsymbol{\Psi}) = H(S) - H(S|\boldsymbol{\Phi}, \boldsymbol{\Psi}) = H(S) \,, \tag{3.8}$$

thus measurement of the state of the universe fully determines the trajectory outcome and origin.

However, the microstate of the environment is not typically resolved, instead its interaction with the system is coarse-grained into friction and fluctuations [1]. Measurement of the system state alone does not fully determine the trajectory outcome and origin, which become random variables with positive conditional Shannon entropy

$$H(S|\boldsymbol{\Phi}) \equiv -\sum_{s,\boldsymbol{\phi}} p(\boldsymbol{\phi}, s) \ln p(s|\boldsymbol{\phi}) > 0 \tag{3.9}$$

reflecting uncertainty in the state of the environment that is relevant to classification of the current subensemble.

3.2 Joint System and Subensemble Dynamics and Irreversible Entropy Production

The main utility of introducing random variables for the trajectory outcome and origin is that reactive events are now part of an ergodic trajectory that unidirectionally cycles through subensembles and forms a nonequilibrium steady state, allowing the use of tools and concepts from stochastic thermodynamics outlined in Sect. 2.1.3 to discuss the statistics of reactive trajectories. In particular, the entropy production of the transition-path ensemble is now derived and its equivalence with information generated about the trajectory outcome S_+ is proved.

The joint dynamics of $(\boldsymbol{\Phi}, \boldsymbol{S})$ are described using the Master equation

$$\mathrm{d}_t p(\boldsymbol{\phi}, s) = \sum_{\boldsymbol{\phi}', s'} T^{ss'}_{\boldsymbol{\phi}\boldsymbol{\phi}'} \, p(\boldsymbol{\phi}', s') \tag{3.10}$$

where $T^{ss'}_{\boldsymbol{\phi}\boldsymbol{\phi}'}$ is the $(\boldsymbol{\phi}', s') \to (\boldsymbol{\phi}, s)$ transition rate. Conditional transition rates for the system in the $A \to B$ subensemble were derived in Ref. [3], and they are adapted for all subensembles below.

Following [3], the transition probability over time $\mathrm{d}t$ for a $\boldsymbol{\phi}' \to \boldsymbol{\phi}$ transition given the trajectory remains in subensemble $\boldsymbol{s}$ is

$$T^{s=s'}_{\boldsymbol{\phi}\boldsymbol{\phi}'} \mathrm{d}t = p(\boldsymbol{\phi}, s_+, s_- | \boldsymbol{\phi}', s'_+, s'_-) \tag{3.11a}$$

$$= p(s_+, s_- | \boldsymbol{\phi}, \boldsymbol{\phi}', s'_+, s'_-) \, p(\boldsymbol{\phi}|\boldsymbol{\phi}', s'_+, s'_-) \tag{3.11b}$$

$$= p(\boldsymbol{\phi}|\boldsymbol{\phi}', s'_+, s'_-) \tag{3.11c}$$

$$= p(\boldsymbol{\phi}|\boldsymbol{\phi}', s'_+) \tag{3.11d}$$

$$= \frac{p(\boldsymbol{\phi}, s'_+|\boldsymbol{\phi}')}{p(s'_+|\boldsymbol{\phi}')} \tag{3.11e}$$

$$= \frac{p(s'_+|\boldsymbol{\phi}) \, p(\boldsymbol{\phi}|\boldsymbol{\phi}')}{p(s'_+|\boldsymbol{\phi}')} \, . \tag{3.11f}$$

Equation (3.11b) splits the joint probability into conditional and marginal probabilities. In (3.11c), $p(s_+, s_- | \boldsymbol{\phi}, \boldsymbol{\phi}', s'_+, s'_-) = 1$ when the subensemble does not change. In (3.11d), the Markov property is used to eliminate the dependence of the next state on trajectory origin. In (3.11e), the conditional probability is expressed as the ratio of joint and marginal probabilities. In (3.11f), it is recognized that the trajectory outcome depends only on the most recent state $\boldsymbol{\phi}$ and drop the dependency on $\boldsymbol{\phi}'$. Finally, recalling that $s_+ = s'_+$ for these transitions and re-expressing the transition probability as a transition rate, the rate of a $\boldsymbol{\phi}' \to \boldsymbol{\phi}$ transition is subensemble $\boldsymbol{s}$ is

$$T^{s=s'}_{\boldsymbol{\phi}\boldsymbol{\phi}'} = \frac{p(s_+|\boldsymbol{\phi})}{p(s_+|\boldsymbol{\phi}')} T_{\boldsymbol{\phi}\boldsymbol{\phi}'} \,. \tag{3.12}$$

Transition rates for the four sets of transitions that change subensemble are similarly derived, where either the trajectory origin or outcome changes while the other is constant. The transition probability in time $\mathrm{d}t$ for a $\boldsymbol{\phi}' \to \boldsymbol{\phi}$ transition out of A where the subensemble changes from $\boldsymbol{s}' = (A, A) \to \boldsymbol{s} = (A, B)$ is:

$$T^{s=(A,B),s'=(A,A)}_{\boldsymbol{\phi}\boldsymbol{\phi}'} \mathrm{d}t = p(\boldsymbol{\phi}, S_+ = B, S_- = A | \boldsymbol{\phi}', S'_+ = A, S'_- = A) \tag{3.13a}$$

$$= p(S_+ = B, S_- = A | \boldsymbol{\phi}, \boldsymbol{\phi}', S'_+ = A, S'_- = A)$$
$$p(\boldsymbol{\phi}|\boldsymbol{\phi}', S'_+ = A, S'_- = A) \tag{3.13b}$$

$$= p(S_+ = B | \boldsymbol{\phi}, \boldsymbol{\phi}', S'_+ = A, S'_- = A)$$
$$p(\boldsymbol{\phi}|\boldsymbol{\phi}', S'_+ = A, S'_- = A) \tag{3.13c}$$

$$= p(S_+ = B|\boldsymbol{\phi})\; p(\boldsymbol{\phi}|\boldsymbol{\phi}') \,. \tag{3.13d}$$

(3.13c) uses the fact that $S_- = A$ when $\boldsymbol{\phi}' \in A$. (3.13d) uses the Markov property to simplify the conditional distributions. The transition probability is then re-expressed as a transition rate

$$T^{s=(A,B),s'=(A,A)}_{\boldsymbol{\phi}\boldsymbol{\phi}'} = p(S_+ = B|\boldsymbol{\phi})\, T_{\boldsymbol{\phi}\boldsymbol{\phi}'} \,. \tag{3.14}$$

Similarly, the transition rate for a $\boldsymbol{\phi}' \to \boldsymbol{\phi}$ transition where the subensemble outcome changes from $\boldsymbol{s}' = (B, B) \to \boldsymbol{s}' = (B, A)$ is

$$T^{s=(B,A),s'=(B,B)}_{\boldsymbol{\phi}\boldsymbol{\phi}'} = p(S_+ = A|\boldsymbol{\phi})\, T_{\boldsymbol{\phi}\boldsymbol{\phi}'} \,. \tag{3.15}$$

The trajectory origin changes when the system finishes a transition path at the boundary of A or B. The probability for a $\boldsymbol{\phi}' \to \boldsymbol{\phi}$ transition into A where the subensemble changes from $\boldsymbol{s}' = (B, A) \to \boldsymbol{s} = (A, A)$ is:

$$T^{s=(A,A),s'=(B,A)}_{\boldsymbol{\phi}\boldsymbol{\phi}'} \mathrm{d}t = p(\boldsymbol{\phi}, S_+ = A, S_- = A | \boldsymbol{\phi}', S'_+ = A, S'_- = B) \tag{3.16a}$$

$$= p(S_+ = A, S_- = A|\boldsymbol{\phi}, \boldsymbol{\phi}', S'_+ = A, S'_- = B)$$

$$p(\boldsymbol{\phi}|\boldsymbol{\phi}', S'_+ = A, S'_- = B) \tag{3.16b}$$

$$= p(\boldsymbol{\phi}|\boldsymbol{\phi}', S'_+ = A) \tag{3.16c}$$

$$= \frac{p(\boldsymbol{\phi}, S'_+ = A|\boldsymbol{\phi}')}{p(S'_+ = A|\boldsymbol{\phi}')} \tag{3.16d}$$

$$= \frac{p(S'_+ = A|\boldsymbol{\phi}, \boldsymbol{\phi}')\, p(\boldsymbol{\phi}|\boldsymbol{\phi}')}{p(S'_+ = A|\boldsymbol{\phi}')} \tag{3.16e}$$

$$= \frac{p(\boldsymbol{\phi}|\boldsymbol{\phi}')}{p(S'_+ = A|\boldsymbol{\phi}')} \,. \tag{3.16f}$$

(3.16c) recognizes that $S_+ = A$ and $S_- = A$ for $\boldsymbol{\phi} \in A$ and uses the Markov property to eliminate dependence on S'_- in $p(\boldsymbol{\phi}|\boldsymbol{\phi}', S'_+ = A, S'_- = A)$. (3.16f) uses $S'_+ = A$ for $\boldsymbol{\phi} \in A$. Finally, the transition probability is re-expressed as the transition rate

$$T^{s=(A,A),s'=(B,A)}_{\boldsymbol{\phi}\boldsymbol{\phi}'} = \frac{1}{p(S_+ = A|\boldsymbol{\phi}')} T_{\boldsymbol{\phi}\boldsymbol{\phi}'} \,, \tag{3.17}$$

and similarly derive the transition rate for a $\boldsymbol{\phi}' \to \boldsymbol{\phi}$ transition where the system enters B and finishes a forward TPE ($s' = (A, B) \to s = (B, B)$) as

$$T^{s=(B,B),s'=(A,B)}_{\boldsymbol{\phi}\boldsymbol{\phi}'} = \frac{1}{p(S_+ = B|\boldsymbol{\phi}')} T_{\boldsymbol{\phi}\boldsymbol{\phi}'} \,. \tag{3.18}$$

When averaged over all such transitions, these transition rates yield the unidirectional probability flux ν_S between trajectory subensembles (3.4),

$$\nu_S = \sum_{\boldsymbol{\phi}\notin A, \boldsymbol{\phi}'\in A} p(S_+ = B|\boldsymbol{\phi})\, T_{\boldsymbol{\phi}\boldsymbol{\phi}'}\pi(\boldsymbol{\phi}') \tag{3.19a}$$

$$= \sum_{\boldsymbol{\phi}\notin B, \boldsymbol{\phi}'\in B} p(S_+ = A|\boldsymbol{\phi})\, T_{\boldsymbol{\phi}\boldsymbol{\phi}'}\pi(\boldsymbol{\phi}') \tag{3.19b}$$

$$= \sum_{\boldsymbol{\phi}\in A, \boldsymbol{\phi}'\notin A} T_{\boldsymbol{\phi}\boldsymbol{\phi}'}\pi(\boldsymbol{\phi}')\, p(S_- = B|\boldsymbol{\phi}') \tag{3.19c}$$

$$= \sum_{\boldsymbol{\phi}\in B, \boldsymbol{\phi}'\notin B} T_{\boldsymbol{\phi}\boldsymbol{\phi}'}\pi(\boldsymbol{\phi}')\, p(S_- = A|\boldsymbol{\phi}') \,. \tag{3.19d}$$

Note that each RHS of (3.19) has an implicit conditional probability for the other element of the trajectory subensemble, each of which equals unity on the relevant system subspace, e.g., $p(S_- = A|\boldsymbol{\phi}') = 1$ for $\boldsymbol{\phi}' \in A$ in (3.19a).

To summarize, the transition rates for joint system-subensemble dynamics are:

$$T^{ss'}_{\phi\phi'} = \begin{cases} T^{s}_{\phi\phi'} \equiv T_{\phi\phi'} \frac{p(s_+|\phi)}{p(s_+|\phi')}, & s' = s \\ T_{\phi\phi'} p(S_+ = B|\phi), & \begin{cases} \phi' \in A, \phi \notin A, \\ s' = (A, A), s = (A, B) \end{cases} \\ T_{\phi\phi'} p(S_+ = A|\phi), & \begin{cases} \phi' \in B, \phi \notin B, \\ s' = (B, B), s = (B, A) \end{cases} \\ T_{\phi\phi'}/p(S_+ = A|\phi'), & \begin{cases} \phi' \notin A, \phi \in A, \\ s' = (B, A), s = (A, A) \end{cases} \\ T_{\phi\phi'}/p(S_+ = B|\phi'), & \begin{cases} \phi' \notin B, \phi \in B, \\ s' = (A, B), s = (B, B) \end{cases} \\ -\sum_{\substack{\phi''\neq\phi' \\ s''\neq s'}} T^{s''s'}_{\phi''\phi'}, & \phi = \phi', s = s' \\ 0 & \text{otherwise} \end{cases} . \tag{3.20}$$

The top transition does not change the subensemble, and biases transitions within subensemble s toward states with higher probability of trajectory outcome s_+. The middle four transitions switch subensembles and are unidirectional, contributing to the probability flux ν_S (Eqs. (3.3), (3.19)). The $(\phi', s') \to (\phi, s)$ transition rate is defined such that the probability distribution remains normalized under time evolution (Eq. (2.9)). These joint dynamics are Markovian: since the underlying system dynamics are Markovian, the transition rates (3.20) do not depend on the trajectory origin s_-, and the outcome s_+ does not induce dependence on earlier system states. Considered alone, system dynamics are at equilibrium and microscopically reversible; adding the trajectory outcome and origin variables (that are not functions of system state and explicitly depend on the past and future) breaks time-reversal symmetry, producing absolutely irreversible trajectory-subensemble transitions and time-asymmetric system transitions within a given subensemble.

3.2.1 *Time-Asymmetry of Transition-Path Dynamics*

To quantify the time-asymmetry for a particular $\phi' \to \phi$ transition in subensemble s, (3.20), Bayes' rule, and the equilibrium detailed-balance relation $T_{\phi\phi'}\pi(\phi') = T_{\phi'\phi}\pi(\phi)$ are combined to derive a local detailed-balance relation,

$$\frac{T^s_{\phi\phi'}p(\phi'|s)}{T^s_{\phi'\phi}p(\phi|s)} = \frac{p(s_-|\phi')p(s_+|\phi)}{p(s_-|\phi)p(s_+|\phi')} . \tag{3.21}$$

The $A \to A$ (and analogously $B \to B$) stationary subensemble has $s_+ = s_- = A$, and due to system detailed balance $p(S_- = A|\phi) = p(S_+ = A|\phi)$, so the RHS is unity and detailed balance holds for transitions within stationary subensembles. The reactive subensembles (forward or reverse TPE) have different trajectory outcome and origin so the RHS side differs from unity, leading to a detailed-balance-breaking flux (and hence entropy production) along particular transitions within these subensembles.

Within a fixed subensemble s, the net trajectory flux is

$$J^s_{\phi\phi'} = T^s_{\phi\phi'}p(\phi', s) - T^s_{\phi'\phi}p(\phi, s) \tag{3.22a}$$

$$= \left[p(s_-|\phi')p(s_+|\phi) - p(s_-|\phi)p(s_+|\phi')\right]T_{\phi\phi'}\pi(\phi') . \tag{3.22b}$$

The second equality follows from $p(\phi, s) = p(s|\phi)\pi(\phi)$; the conditional independence of s_+ and s_- given state ϕ, i.e., $p(s_+, s_-|\phi) = p(s_+|\phi)p(s_-|\phi)$; and substitution for $T^s_{\phi'\phi}$ using (3.20). The stationary subensembles ($A \to A$ and $B \to B$) have no net flux because each trajectory segment and its time-reversed counterpart occur at equal rates within the same subensemble. In contrast, the forward and reverse TPEs have net trajectory flux since each transition path and its time-reversed counterpart occur in different subensembles.

3.2.2 Entropy Production for Joint Dynamics

Time-asymmetry within the transition-path ensembles implies that there is irreversible entropy production for dynamics in these subensembles. The change in joint entropy $H(\Phi, S) \equiv -\sum_{\phi,s} p(\phi, s) \ln p(\phi, s)$ is decomposed into three terms [7]

$$0 = \mathrm{d}_t H(\Phi, S) = \underbrace{\sum_{\phi,\phi',s} T^s_{\phi\phi'}p(\phi', s) \ln \frac{T^s_{\phi\phi'}p(\phi', s)}{T^s_{\phi'\phi}p(\phi, s)}}_{\dot{H}^{\mathrm{irr}}(\Phi,S)} - \underbrace{\sum_{\phi,\phi',s} T^s_{\phi\phi'}p(\phi', s) \ln \frac{T^s_{\phi\phi'}}{T^s_{\phi'\phi}}}_{\dot{H}^{\mathrm{env}}(\Phi,S)}$$

$$+ \underbrace{\sum_{\phi,\phi',s\neq s'} T^{ss'}_{\phi\phi'}p(\phi', s') \ln \frac{p(\phi', s')}{p(\phi, s)}}_{\dot{H}^{\mathrm{sub}}(\Phi,S)} , \tag{3.23}$$

where $\dot{H}^{\text{irr}}(\boldsymbol{\Phi}, \boldsymbol{S})$ is the irreversible entropy production and $\dot{H}^{\text{env}}(\boldsymbol{\Phi}, \boldsymbol{S})$ the environmental entropy change for transitions that do not change the subensemble, and $\dot{H}^{\text{sub}}(\boldsymbol{\Phi}, \boldsymbol{S})$ is the change in joint entropy due to transitions that change the subensemble.

To simplify the entropy change for transitions that change the subensemble, the joint transition rates (Eq. (3.20)) for these four sets of transitions are inserted into the third term of Eq. (3.23) to obtain:

$$
\begin{aligned}
\dot{H}^{\text{sub}}(\boldsymbol{\Phi}, \boldsymbol{S}) = & \sum_{\boldsymbol{\phi} \in A, \boldsymbol{\phi}' \notin A} T_{\boldsymbol{\phi}\boldsymbol{\phi}'} \pi(\boldsymbol{\phi}') p(S_- = B|\boldsymbol{\phi}') \ln \frac{p(\boldsymbol{\phi}', \boldsymbol{S} = (B, A))}{p(\boldsymbol{\phi}, \boldsymbol{S} = (A, A))} \\
& + \sum_{\boldsymbol{\phi} \notin A, \boldsymbol{\phi}' \in A} T_{\boldsymbol{\phi}\boldsymbol{\phi}'} \pi(\boldsymbol{\phi}') p(S_+ = B|\boldsymbol{\phi}) \ln \frac{p(\boldsymbol{\phi}', \boldsymbol{S} = (A, A))}{p(\boldsymbol{\phi}, \boldsymbol{S} = (A, B))} \\
& + \sum_{\boldsymbol{\phi} \in B, \boldsymbol{\phi}' \notin B} T_{\boldsymbol{\phi}\boldsymbol{\phi}'} \pi(\boldsymbol{\phi}') p(S_- = A|\boldsymbol{\phi}') \ln \frac{p(\boldsymbol{\phi}', \boldsymbol{S} = (A, B))}{p(\boldsymbol{\phi}, \boldsymbol{S} = (B, B))} \\
& + \sum_{\boldsymbol{\phi} \notin B, \boldsymbol{\phi}' \in B} T_{\boldsymbol{\phi}\boldsymbol{\phi}'} \pi(\boldsymbol{\phi}') p(S_+ = A|\boldsymbol{\phi}) \ln \frac{p(\boldsymbol{\phi}', \boldsymbol{S} = (B, B))}{p(\boldsymbol{\phi}, \boldsymbol{S} = (B, A))}
\end{aligned}
\tag{3.24a}
$$

$$
\begin{aligned}
= & \sum_{\boldsymbol{\phi} \notin A, \boldsymbol{\phi}' \in A} T_{\boldsymbol{\phi}'\boldsymbol{\phi}} \pi(\boldsymbol{\phi}) p(S_- = B|\boldsymbol{\phi}) \ln \frac{p(\boldsymbol{\phi}, \boldsymbol{S} = (B, A))}{\pi(\boldsymbol{\phi}')} \\
& - \sum_{\boldsymbol{\phi} \notin A, \boldsymbol{\phi}' \in A} T_{\boldsymbol{\phi}\boldsymbol{\phi}'} \pi(\boldsymbol{\phi}') p(S_+ = B|\boldsymbol{\phi}) \ln \frac{p(\boldsymbol{\phi}, \boldsymbol{S} = (A, B))}{\pi(\boldsymbol{\phi}')} \\
& + \sum_{\boldsymbol{\phi} \notin B, \boldsymbol{\phi}' \in B} T_{\boldsymbol{\phi}'\boldsymbol{\phi}} \pi(\boldsymbol{\phi}) p(S_- = A|\boldsymbol{\phi}) \ln \frac{p(\boldsymbol{\phi}, \boldsymbol{S} = (A, B))}{\pi(\boldsymbol{\phi}')} \\
& - \sum_{\boldsymbol{\phi} \notin B, \boldsymbol{\phi}' \in B} T_{\boldsymbol{\phi}\boldsymbol{\phi}'} \pi(\boldsymbol{\phi}') p(S_+ = A|\boldsymbol{\phi}) \ln \frac{p(\boldsymbol{\phi}, \boldsymbol{S} = (B, A))}{\pi(\boldsymbol{\phi}')}
\end{aligned}
\tag{3.24b}
$$

$$
= 0\,. \tag{3.24c}
$$

In (3.24a), the indices in the first and third terms are rewritten, and the simplifications $p(\boldsymbol{\phi}, \boldsymbol{S} = (A, A)) = \pi(\boldsymbol{\phi})$ for $\boldsymbol{\phi} \in A$ and $p(\boldsymbol{\phi}, \boldsymbol{S} = (B, B)) = \pi(\boldsymbol{\phi})$ for $\boldsymbol{\phi} \in B$ are used. The first and second terms and the third and fourth terms cancel because of the equilibrium relationships $p(S_+ = A|\boldsymbol{\phi}) = p(S_- = A|\boldsymbol{\phi})$, $p(S_+ = B|\boldsymbol{\phi}) = p(S_- = B|\boldsymbol{\phi})$, $p(\phi, \boldsymbol{S} = (B, A)) = p(\phi, \boldsymbol{S} = (A, B))$, and $T_{\boldsymbol{\phi}\boldsymbol{\phi}'} \pi(\boldsymbol{\phi}') = T_{\boldsymbol{\phi}'\boldsymbol{\phi}} \pi(\boldsymbol{\phi})$. Therefore subensemble transitions do not change the joint entropy.

Substituting in the joint transition rates (3.20) for dynamics within a subensemble $T^s_{\phi\phi'} = T_{\phi\phi'} p(s_+|\phi)/p(s_+|\phi')$, the environmental entropy change is

$$\dot{H}^{\text{env}}(\mathbf{\Phi}, \mathbf{S}) = \sum_{\phi,\phi',s} T^s_{\phi\phi'} p(\phi', s) \left[\ln \frac{T_{\phi\phi'}}{T_{\phi'\phi}} + 2 \ln \frac{p(s_+|\phi)}{p(s_+|\phi')} \right] \tag{3.25a}$$

$$= 2 \sum_{\phi,\phi',s} T^s_{\phi\phi'} p(\phi', s) \ln \frac{p(s_+|\phi)}{p(s_+|\phi')} \tag{3.25b}$$

$$= 2 \sum_{\phi,\phi',s_+} T^{s_+}_{\phi\phi'} p(\phi', s_+) \ln \frac{p(s_+|\phi)}{p(s_+|\phi')} \tag{3.25c}$$

$$= 2\dot{I}^{\mathbf{\Phi}}(S_+; \mathbf{\Phi}) , \tag{3.25d}$$

where $\dot{I}^{\mathbf{\Phi}}(S_+; \mathbf{\Phi}) \geq 0$ is the rate of change in mutual information between the trajectory outcome and system state due to system dynamics in a fixed subensemble [8]. Equation (3.25b) is obtained by summing over s to cancel contributions from the equilibrium transition rates, and (3.25c) is obtained by summing over the trajectory origin S_-.

Finally, the irreversible entropy production $\dot{H}^{\text{irr}}(\mathbf{\Phi}, \mathbf{S})$ of the cycle is expressed in terms of the irreversible entropy production of dynamics given fixed subensemble s, $\dot{\Sigma}_s$, which quantifies the time irreversibility of system dynamics *within* that subensemble [9]:

$$\dot{H}^{\text{irr}}(\mathbf{\Phi}, \mathbf{S}) = \sum_s p(s) \sum_{\phi,\phi'} T^s_{\phi\phi'} p(\phi'|s) \ln \frac{T^s_{\phi\phi'} p(\phi'|s)}{T^s_{\phi'\phi} p(\phi|s)} \tag{3.26a}$$

$$= \sum_s p(s) \dot{\Sigma}_s \tag{3.26b}$$

$$\equiv \langle \dot{\Sigma} \rangle , \tag{3.26c}$$

where $\langle \dot{\Sigma} \rangle \equiv \sum_s p(s) \dot{\Sigma}_s \geq 0$ is the subensemble-weighted average of the irreversible entropy production rate within each subensemble.

Rearranging (3.23) gives:

$$0 \leq \langle \dot{\Sigma} \rangle = 2\dot{I}^{\mathbf{\Phi}}(S_+; \mathbf{\Phi}) \tag{3.27a}$$

$$= \dot{I}^{\mathbf{\Phi}}(S_+; \mathbf{\Phi}) - \dot{I}^{\mathbf{\Phi}}(S_-; \mathbf{\Phi}) , \tag{3.27b}$$

where $\dot{I}^{\mathbf{\Phi}}(S_-; \mathbf{\Phi}) \leq 0$ is the rate of change in mutual information between trajectory origin and system state due to $\mathbf{\Phi}$ dynamics in a fixed subensemble. The rate of change in mutual information between system state and trajectory *outcome* due to $\mathbf{\Phi}$ dynamics is of equal magnitude and opposite sign from the rate of change in mutual information between system state and trajectory *origin* due to $\mathbf{\Phi}$ dynamics:

$$\dot{\imath}^{\mathbf{\Phi}}(S_+;\mathbf{\Phi}) = \sum_{\boldsymbol{\phi},\boldsymbol{\phi}',s_+} T^{s_+}_{\boldsymbol{\phi}\boldsymbol{\phi}'}\, p(\boldsymbol{\phi}',s_+) \ln \frac{p(s_+|\boldsymbol{\phi})}{p(s_+|\boldsymbol{\phi}')} \tag{3.28a}$$

$$= \sum_{\boldsymbol{\phi},\boldsymbol{\phi}',s_+} T_{\boldsymbol{\phi}\boldsymbol{\phi}'}\pi(\boldsymbol{\phi}')p(s_+|\boldsymbol{\phi}) \ln \frac{p(s_+|\boldsymbol{\phi})}{p(s_+|\boldsymbol{\phi}')} \tag{3.28b}$$

$$= \sum_{\boldsymbol{\phi},\boldsymbol{\phi}',s_-} T_{\boldsymbol{\phi}\boldsymbol{\phi}'}\pi(\boldsymbol{\phi}')p(s_-|\boldsymbol{\phi}) \ln \frac{p(s_-|\boldsymbol{\phi})}{p(s_-|\boldsymbol{\phi}')} \tag{3.28c}$$

$$= \sum_{\boldsymbol{\phi},\boldsymbol{\phi}',s_-} T_{\boldsymbol{\phi}'\boldsymbol{\phi}}\pi(\boldsymbol{\phi})p(s_-|\boldsymbol{\phi}) \ln \frac{p(s_-|\boldsymbol{\phi})}{p(s_-|\boldsymbol{\phi}')} \tag{3.28d}$$

$$= -\sum_{\boldsymbol{\phi},\boldsymbol{\phi}',s_-} T_{\boldsymbol{\phi}'\boldsymbol{\phi}}\, p(\boldsymbol{\phi},s_-) \ln \frac{p(s_-|\boldsymbol{\phi}')}{p(s_-|\boldsymbol{\phi})} \tag{3.28e}$$

$$= -\dot{\imath}^{\mathbf{\Phi}}(S_-;\mathbf{\Phi})\,. \tag{3.28f}$$

In (3.28b) the joint transition rate is expanded using the subensemble-specific transition rates (3.20); (3.28c) relates outcome and origin probabilities using equilibrium relations $p(S_+ = A|\boldsymbol{\phi}) = p(S_- = A|\boldsymbol{\phi})$, $p(S_+ = B|\boldsymbol{\phi}) = p(S_- = B|\boldsymbol{\phi})$; and (3.28d) uses detailed balance, $T_{\boldsymbol{\phi}\boldsymbol{\phi}'}\pi(\boldsymbol{\phi}') = T_{\boldsymbol{\phi}'\boldsymbol{\phi}}\pi(\boldsymbol{\phi})$.

These information rates reflect the dependence of the trajectory outcome and origin variables on the past and future states of the system: as the system evolves, uncertainty about the outcome S_+ diminishes, increasing the information the current system state carries about S_+, while uncertainty (given current system state $\mathbf{\Phi}$) about the origin S_- increases, decreasing information $\mathbf{\Phi}$ carries about S_-.

Since the stationary subensembles have no net flux, they also have no entropy production ($\dot{\Sigma}_{s=(A,A)} = \dot{\Sigma}_{s=(B,B)} = 0$), hence (3.27a) reduces to an equation for a single subensemble,

$$0 \leq p_{\mathrm{R}}\dot{\Sigma}_{\mathrm{R}} = \dot{\imath}^{\mathbf{\Phi}}(S_+;\mathbf{\Phi})\,. \tag{3.29}$$

This equates the rate $\dot{\imath}^{\mathbf{\Phi}}(S_+;\mathbf{\Phi})$ of generating information about the outcome with the product of the entropy production rate of a reactive subensemble $\dot{\Sigma}_{\mathrm{R}} = \dot{\Sigma}_{s=(A,B)} = \dot{\Sigma}_{s=(B,A)}$ and that subensemble's marginal probability $p_{\mathrm{R}} = p(S = (A,B)) = p(S = (B,A))$. Although the supertrajectory is at equilibrium with no entropy production, $\dot{\Sigma}_{\mathrm{R}}$ physically represents the dissipation that would be necessary in a system evolving according to the TPE's detailed-balance-breaking transition rates (top line of (3.20) for $s_- \neq s_+$). Equation (3.29) is the second major result: the entropy production in a reactive subensemble equals the information generated about the reactivity of trajectories.

3.2.3 Thermodynamic Metric

The TPE entropy production (3.29) can also be cast geometrically, by deriving a Fisher-information metric $\mathcal{I}(\boldsymbol{\phi})$ that imposes an information geometry on the state space [10, 11]. Suppose the state space $\boldsymbol{\Phi}$ is continuous and represented by a set of coordinates $\{\boldsymbol{\phi}_i\}$. The Master equation $\mathrm{d}_t p(\boldsymbol{\phi}) = \sum_{\boldsymbol{\phi}'} T_{\boldsymbol{\phi}\boldsymbol{\phi}'} p(\boldsymbol{\phi}')$ represents a discrete approximation of the continuous dynamics [1]. The difference in mutual information rates (3.27b) is rearranged to obtain the transition-weighted relative entropy [6]

$$D[p(s|\boldsymbol{\phi}')||p(s|\boldsymbol{\phi})] \equiv \sum_s p(s|\boldsymbol{\phi}') \ln \frac{p(s|\boldsymbol{\phi}')}{p(s|\boldsymbol{\phi})} \tag{3.30}$$

between the conditional subensemble distributions $p(s|\boldsymbol{\phi}')$ and $p(s|\boldsymbol{\phi})$ before and after the transition, respectively, then expand in small state changes $\boldsymbol{\phi} - \boldsymbol{\phi}'$:

$$p_{\mathrm{R}} \dot{\Sigma}_{\mathrm{R}} = \dot{I}^{\Phi}(S_+; \Phi) - \dot{I}^{\Phi}(S_-; \Phi) \tag{3.31a}$$

$$= \sum_{\boldsymbol{\phi},\boldsymbol{\phi}',s_+} T_{\boldsymbol{\phi}\boldsymbol{\phi}'} \pi(\boldsymbol{\phi}') p(s_+|\boldsymbol{\phi}) \ln \frac{p(s_+|\boldsymbol{\phi})}{p(s_+|\boldsymbol{\phi}')} - \sum_{\boldsymbol{\phi},\boldsymbol{\phi}',s_-} T_{\boldsymbol{\phi}'\boldsymbol{\phi}} \pi(\boldsymbol{\phi}) p(s_-|\boldsymbol{\phi}) \ln \frac{p(s_-|\boldsymbol{\phi}')}{p(s_-|\boldsymbol{\phi})} \tag{3.31b}$$

$$= \sum_{\boldsymbol{\phi},\boldsymbol{\phi}',s_+,s_-} T_{\boldsymbol{\phi}\boldsymbol{\phi}'} \pi(\boldsymbol{\phi}') p(s_+|\boldsymbol{\phi}) p(s_-|\boldsymbol{\phi}) \ln \frac{p(s_+|\boldsymbol{\phi})}{p(s_+|\boldsymbol{\phi}')} - \sum_{\boldsymbol{\phi},\boldsymbol{\phi}',s_+,s_-} T_{\boldsymbol{\phi}'\boldsymbol{\phi}} \pi(\boldsymbol{\phi}) p(s_-|\boldsymbol{\phi}) p(s_+|\boldsymbol{\phi}) \ln \frac{p(s_-|\boldsymbol{\phi}')}{p(s_-|\boldsymbol{\phi})} \tag{3.31c}$$

$$= \sum_{\boldsymbol{\phi},\boldsymbol{\phi}',s_+,s_-} T_{\boldsymbol{\phi}\boldsymbol{\phi}'} \pi(\boldsymbol{\phi}') p(s_+|\boldsymbol{\phi}) p(s_-|\boldsymbol{\phi}) \ln \frac{p(s_+|\boldsymbol{\phi}) p(s_-|\boldsymbol{\phi})}{p(s_+|\boldsymbol{\phi}') p(s_-|\boldsymbol{\phi}')} \tag{3.31d}$$

$$= \sum_{\boldsymbol{\phi},\boldsymbol{\phi}'} T_{\boldsymbol{\phi}\boldsymbol{\phi}'} \pi(\boldsymbol{\phi}') D[p(s|\boldsymbol{\phi}')||p(s|\boldsymbol{\phi})] \tag{3.31e}$$

$$\approx \sum_{\boldsymbol{\phi},\boldsymbol{\phi}'} T_{\boldsymbol{\phi}\boldsymbol{\phi}'} \pi(\boldsymbol{\phi}') \tfrac{1}{2} \sum_{i,j} (\phi_i - \phi_i') \mathcal{I}_{ij}(\boldsymbol{\phi}') (\phi_j - \phi_j') \,. \tag{3.31f}$$

In (3.31c), each term is multiplied by unity ($\sum_{s_-} p(s_-|\boldsymbol{\phi})$ and $\sum_{s_+} p(s_+|\boldsymbol{\phi})$ respectively), then detailed balance ($T_{\boldsymbol{\phi}\boldsymbol{\phi}'} \pi(\boldsymbol{\phi}') = T_{\boldsymbol{\phi}'\boldsymbol{\phi}} \pi(\boldsymbol{\phi})$) is used to combine terms in (3.31d). The ith component of the state-space vector is denoted by $\boldsymbol{\phi}_i$, and $\mathcal{I}_{ij}(\boldsymbol{\phi})$ is the Fisher information of the trajectory outcome/origin distribution at state $\boldsymbol{\phi}$,

$$\mathcal{I}_{ij}(\boldsymbol{\phi}) \equiv \sum_{s} p\left(s|\boldsymbol{\phi}\right) \frac{\partial \ln p(s|\boldsymbol{\phi})}{\partial \boldsymbol{\phi}_i} \frac{\partial \ln p(s|\boldsymbol{\phi})}{\partial \boldsymbol{\phi}_j} \tag{3.32a}$$

$$= \sum_{s} \frac{1}{p\left(s|\boldsymbol{\phi}\right)} \frac{\partial p(s|\boldsymbol{\phi})}{\partial \boldsymbol{\phi}_i} \frac{\partial p(s|\boldsymbol{\phi})}{\partial \boldsymbol{\phi}_j} \tag{3.32b}$$

$$= \frac{2}{\left(1-q^{+}(\boldsymbol{\phi})\right)q^{+}(\boldsymbol{\phi})} \frac{\partial\left(1-q^{+}(\boldsymbol{\phi})\right)q^{+}(\boldsymbol{\phi})}{\partial \boldsymbol{\phi}_i} \frac{\partial\left(1-q^{+}(\boldsymbol{\phi})\right)q^{+}(\boldsymbol{\phi})}{\partial \boldsymbol{\phi}_j} \tag{3.32c}$$

$$+ \frac{1}{\left(1-q^{+}(\boldsymbol{\phi})\right)^2} \frac{\partial\left(1-q^{+}(\boldsymbol{\phi})\right)^2}{\partial \boldsymbol{\phi}_i} \frac{\partial\left(1-q^{+}(\boldsymbol{\phi})\right)^2}{\partial \boldsymbol{\phi}_j}$$

$$+ \frac{1}{\left(q^{+}(\boldsymbol{\phi})\right)^2} \frac{\partial\left(q^{+}(\boldsymbol{\phi})\right)^2}{\partial \boldsymbol{\phi}_i} \frac{\partial\left(q^{+}(\boldsymbol{\phi})\right)^2}{\partial \boldsymbol{\phi}_j}$$

$$= \left[\frac{2\left(1-2q^{+}(\boldsymbol{\phi})\right)^2}{\left(1-q^{+}(\boldsymbol{\phi})\right)q^{+}(\boldsymbol{\phi})} + \frac{4\left(1-q^{+}(\boldsymbol{\phi})\right)^2}{\left(1-q^{+}(\boldsymbol{\phi})\right)^2} + \frac{4\left(q^{+}(\boldsymbol{\phi})\right)^2}{\left(q^{+}(\boldsymbol{\phi})\right)^2}\right] \frac{\partial q^{+}(\boldsymbol{\phi})}{\partial \boldsymbol{\phi}_i} \frac{\partial q^{+}(\boldsymbol{\phi})}{\partial \boldsymbol{\phi}_j} \tag{3.32d}$$

$$= \frac{2}{q^{+}(\boldsymbol{\phi})\left(1-q^{+}(\boldsymbol{\phi})\right)} \frac{\partial q^{+}(\boldsymbol{\phi})}{\partial \boldsymbol{\phi}_i} \frac{\partial q^{+}(\boldsymbol{\phi})}{\partial \boldsymbol{\phi}_j}\,. \tag{3.32e}$$

In (3.32c), each term from (3.32b) is expressed in terms of the forward committor $q^{+}(\boldsymbol{\phi})$. The chain rule is used to pull out a common factor in (3.32d), and the expression is simplified in (3.32e).

In information geometry [10, 11], Fisher information arises as a distance metric in the parameter space of a probability distribution. Here, the conditional probability $p(s|\boldsymbol{\phi})$ changes as the system evolves, where the system state $\boldsymbol{\phi}$ parameterizes the conditional probability distribution through the committor $q^{+}(\boldsymbol{\phi})$. When the system is in A (B), there is no uncertainty about trajectory outcome S_{+} and origin S_{-}, so the conditional probability $p(s|\boldsymbol{\phi} \in A)$ ($p(s|\boldsymbol{\phi} \in B)$) is unity for $s = (A, A)$ ($s = (B, B)$) and zero for all other subensembles. As the system evolves, the conditional distribution over subensembles $p(s|\boldsymbol{\phi})$ corresponding to the current system state changes. The information-geometric distance between successive points in state space where there are changes to coordinates $\boldsymbol{\phi}_i$ and $\boldsymbol{\phi}_j$ is quantified by the Fisher information metric, with squared line element [12]

$$\mathrm{d}\ell^2_{\boldsymbol{\phi}\boldsymbol{\phi}'} \equiv \tfrac{1}{2}\sum_{i,j}(\phi_i - \phi'_i)\mathcal{I}_{ij}(\boldsymbol{\phi}')(\phi_j - \phi'_j)\,. \tag{3.33}$$

Equation (3.31f) is then the rate of mean square distance accumulated by the system evolving at equilibrium,

$$\frac{\langle \mathrm{d}\ell^2 \rangle}{\mathrm{d}t} = \sum_{\boldsymbol{\phi},\boldsymbol{\phi}'} T_{\boldsymbol{\phi}\boldsymbol{\phi}'}\,\pi(\boldsymbol{\phi}')\,\mathrm{d}\ell^2_{\boldsymbol{\phi}\boldsymbol{\phi}'}\,. \tag{3.34}$$

Multiplying by the mean round-trip time $\tau^{\mathrm{RT}} = \tau^{\mathrm{FP}}_{A\to B} + \tau^{\mathrm{FP}}_{B\to A} = (\nu_s)^{-1}$ to complete a cycle through the subensembles (the sum of mean first-passage times from A to B and B to A [4]), the squared reaction-coordinate length $\mathcal{L}^2_{AB}$ is obtained as the mean-squared metric distance for a round trip $A \to B \to A$:

$$\mathcal{L}^2_{AB} \equiv (\tau_A + \tau_B)\frac{\langle \mathrm{d}\ell^2 \rangle}{\mathrm{d}t} \tag{3.35a}$$

$$\approx 2\tau^{\mathrm{RT}} p_{\mathrm{R}}\,\dot{\Sigma}_{\mathrm{R}} \tag{3.35b}$$

$$= 2\tau_{\mathrm{R}}\,\dot{\Sigma}_{\mathrm{R}} \tag{3.35c}$$

$$\dot{\Sigma}_{\mathrm{R}} \approx \frac{\mathcal{L}^2_{AB}}{2\tau_{\mathrm{R}}}\,, \tag{3.35d}$$

for mean transition-path duration $\tau_{\mathrm{R}} = \tau^{\mathrm{RT}} p_{\mathrm{R}}$, given by the mean time to complete a cycle through all subensembles multiplied by the marginal probability of a reactive subensemble [3, 4]. The reaction-coordinate length $\mathcal{L}_{AB}$ roughly quantifies the mean number of fluctuations [13, 14] required for the system to complete a round trip. Therefore, TPE entropy production provides a meaningful measure of the distance between A and B [15], quantifying the time-irreversibility in the transition.

3.3 Quantitative Measure of Coordinate's Relevance to Reaction

The TPE entropy production quantitatively measures the relevance of an arbitrary coordinate to the reaction, which is demonstrated below for a simple model system. The system is assumed to have bipartite dynamics [16, 17], essentially that instantaneous transitions only happen in either a one-dimensional coordinate X or in all other degrees of freedom $\boldsymbol{Y}$ making up the system state $\boldsymbol{\Phi} = (X, \boldsymbol{Y})$:

$$T_{xx',yy'} = \begin{cases} T_{xx',y} & x \neq x' \ , \ y = y' \\ T_{x,yy'} & x = x' \ , \ y \neq y' \\ -\sum\limits_{\substack{x'' \neq x' \\ y'' \neq y'}} T_{x''x',y''y'} & x = x' \ , \ y = y' \ . \\ 0 & \text{otherwise} \end{cases} \tag{3.36}$$

Dynamics that do not obey the bipartite assumption introduce further complications in unambiguously partitioning the entropy production between coordinates [18].

Combining (3.2), (3.22a), and (3.23) gives the full TPE entropy production as a function of the forward committor,

$$p_{\mathrm{R}}\dot{\Sigma}_{\mathrm{R}} = \tfrac{1}{2}\sum_{\boldsymbol{\phi},\boldsymbol{\phi}'} T_{\boldsymbol{\phi}\boldsymbol{\phi}'}\pi(\boldsymbol{\phi}')(q^+(\boldsymbol{\phi}) - q^+(\boldsymbol{\phi}'))\ln\frac{q^+(\boldsymbol{\phi})(1-q^+(\boldsymbol{\phi}'))}{q^+(\boldsymbol{\phi}')(1-q^+(\boldsymbol{\phi}))}\,, \tag{3.37}$$

which splits into contributions from the two transition types:

$$p_{\mathrm{R}}\dot{\Sigma}_{\mathrm{R}} = p_{\mathrm{R}}\dot{\Sigma}_{\mathrm{R}}^{X} + p_{\mathrm{R}}\dot{\Sigma}_{\mathrm{R}}^{Y} \tag{3.38a}$$

$$= \tfrac{1}{2}\sum_{x,x',\mathbf{y}} T_{xx',\mathbf{y}}\pi(x',\mathbf{y})[q^+(x,\mathbf{y}) - q^+(x',\mathbf{y})]\ln\frac{q^+(x',\mathbf{y})[1-q^+(x,\mathbf{y})]}{q^+(x,\mathbf{y})[1-q^+(x',\mathbf{y})]} \tag{3.38b}$$

$$+ \tfrac{1}{2}\sum_{x,\mathbf{y},\mathbf{y}'} T_{x,\mathbf{y}\mathbf{y}'}\pi(x,\mathbf{y}')[q^+(x,\mathbf{y}) - q^+(x,\mathbf{y}')]$$

$$\ln\frac{q^+(x,\mathbf{y})[1-q^+(x,\mathbf{y}')]}{q^+(x,\mathbf{y}')[1-q^+(x,\mathbf{y})]}\,.$$

The same decomposition holds for the information rate in Eq. (3.27a) [8], so that TPE entropy production due to X dynamics is equal to the information rate (due to X dynamics) between $\mathbf{\Phi}$ and S_+:

$$p_{\mathrm{R}}\dot{\Sigma}_{\mathrm{R}}^{X} = \dot{I}^{X}(S_+;\mathbf{\Phi})\,, \tag{3.39}$$

where

$$\dot{I}^{X}(S_+;\mathbf{\Phi}) \equiv \sum_{x,x',\mathbf{y},s_+} T_{xx',\mathbf{y}}^{s_+} p(x',\mathbf{y},s_+)\ln\frac{p(s_+|x,\mathbf{y})}{p(s_+|x',\mathbf{y})}\,. \tag{3.40}$$

This is the third major result: the entropy production due to dynamics of coordinate X equals the mutual information generated by X dynamics, thereby quantifying the relevance of X transitions to identifying the current subensemble and highlighting those transitions that are 'correlated' with reactive trajectories and therefore important to the reaction mechanism.

In particular, for X^* determining the committor and Y^* orthogonal degrees of freedom that are therefore not relevant to the reaction $[q(x, y) = q(x)]$, the entropy production rate due to Y^* dynamics is (simplifying (3.38b)):

$$\dot{\Sigma}_{\mathrm{R}}^{Y^*} = \sum_{x,y,y'} T_{x,yy'}\pi(x, y')[q^+(x) - q^+(x)] \ln \frac{q^+(x)[1 - q^+(x)]}{q^+(x)[1 - q^+(x)]} = 0. \tag{3.41a}$$

Therefore $\dot{\Sigma}_{\mathrm{R}}^{X^*} = \dot{\Sigma}_{\mathrm{R}}$. This is additional confirmation that the committor is the reaction coordinate, in that it provides a thermodynamically complete coarse-grained representation of the transition-path ensemble, fully accounting for its entropy production [9].

3.3.1 Illustrative Example: 2D Bistable Potential

The ability of TPE entropy production to identify relevant reaction coordinates is illustrated with overdamped dynamics in a double-well energy landscape (Fig. 3.2a). The bistable energy potential is separable into terms only depending on the reaction coordinate r and on the bath mode b:

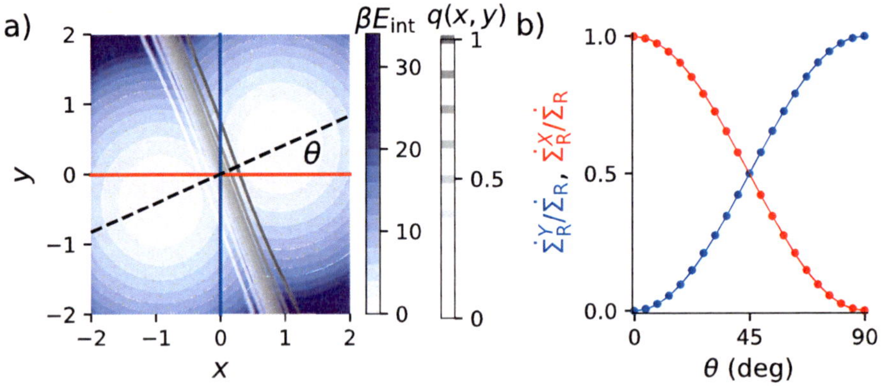

Fig. 3.2 TPE entropy production rate for 2D double-well potential with rotating coordinate system. (**a**) Double-well energy landscape (purple) with fixed x-axis (red solid) and y-axis (blue solid), and rotated reaction coordinate (black dashed). Gray contours: forward committor. (**b**) Share of TPE entropy production due to coordinate X (red dots) and Y (blue dots), as a function of reaction-coordinate angle θ. Red solid: $\cos^2\theta$. Blue solid: $\sin^2\theta$

$$E_{\text{int}}(r, b) = -k_B T \ln\left[e^{-\frac{1}{2}\beta k_m (r+r_m)^2} + e^{-\frac{1}{2}\beta k_m (r-r_m)^2}\right] + \tfrac{1}{2}kb^2\,, \tag{3.42}$$

where $r_m = 1$ defines the locations of the energy minima, and k_m is the landscape curvature near those energy minima, chosen such that the energy barrier $E_{\text{int}}(0, 0) - E_{\text{int}}(r_m, 0)$ is $4k_B T$.

To exemplify the typical situation where the reaction coordinate is not known *a priori* and coordinates are thus chosen based on convenience or intuition, fixed system coordinates (x, y) lie at an angle θ to the correct reaction coordinate, the linear coordinate passing through both energy minima:

$$r = x \cos\theta + y \sin\theta \tag{3.43a}$$

$$b = -x \cos\theta + y \sin\theta\,. \tag{3.43b}$$

The state space (x, y) is divided into discrete segments $\mathrm{d}x = \mathrm{d}y = 0.04 r_m$, and the Master equation with transition rates $T_{xx',yy'} = \Gamma * \min\left[1, e^{-\beta\Delta E_{\text{int}}}\right]$ with diffusion prefactor $\Gamma = 0.1\,\mathrm{d}t^{-1}$ and energy change $\Delta E_{\text{int}} = E_{\text{int}}(x, y) - E_{\text{int}}(x', y')$ [19] is used to evolve the system's dynamics. The committor is solved on the discrete state space using the recursion relations [2]

$$\sum_{\boldsymbol{\phi}'} T_{\boldsymbol{\phi}'\boldsymbol{\phi}}\, q^+(\boldsymbol{\phi}') = 0 \tag{3.44a}$$

$$\sum_{\boldsymbol{\phi}'} T_{\boldsymbol{\phi}\boldsymbol{\phi}'} \frac{\pi(\boldsymbol{\phi}')}{\pi(\boldsymbol{\phi})} q^-(\boldsymbol{\phi}') = 0. \tag{3.44b}$$

Reaction endpoints are defined by $A = \{x, y \,|\, r(x, y) \leq -r_m\}$ and $B = \{x, y \,|\, r(x, y) \geq r_m\}$, so that the committor is independent of b. The TPE entropy production is calculated from (3.38).

For $\theta = 0°$, X is the reaction coordinate, Y is an orthogonal *bath mode* [20], and X dynamics fully capture the TPE entropy production without Y contribution. Figure 3.2b shows that as the underlying energy landscape is rotated relative to system coordinates, the X-coordinate entropy production decreases and Y-coordinate entropy production increases, with equal contribution at $\theta = 45°$. The entropy production for each coordinate is proportional to the squared Euclidean distance between A and B projected onto each coordinate, $\dot{\Sigma}_R^X(\theta) \propto \cos^2\theta$ and $\dot{\Sigma}_R^Y(\theta) \propto \sin^2\theta$. This proportionality is an interesting result in light of the information metric derived in Sect. 3.2.3, and could warrant further investigation into geometric aspects of reaction dynamics and information.

3.4 Discussion

The information thermodynamics of a system undergoing reactions between distinct state-space subsets A and B has been derived, making a fundamental connection between transition-path theory, information theory, and stochastic thermodynamics. Partitioning a long ergodic equilibrium trajectory into reactive and non-reactive subensembles results in entropy production for system dynamics in the reactive subensembles (physically representing the dissipation needed to implement the detailed-balance-breaking transition rates of the TPE), which in turn identifies transitions that are relevant to the overall reaction mechanism. This rigorous equality between TPE entropy production and informativeness of dynamics also holds for an arbitrary coordinate, revealing parallel stochastic-thermodynamic and information-theoretic measures of the relevance of collective variables to the system reaction, that are each maximized by the committor.

This work has implications for the identification of important collective variables and analysis of reaction mechanisms. While the committor provides a microscopically detailed reaction coordinate that maps each system microstate to a scalar value, it does not immediately identify physically meaningful collective variables (e.g., internal molecular coordinates) that are relevant to the reaction [21–23]. These results indicate that relevant coordinates are identified as having nonzero entropy production in the transition-path ensemble; thus partitioning the entropy production between multiple relevant collective variables for that provide physical intuition can yield a low-dimensional model that allows increased insight into the reaction mechanism.

More concretely, this connection that has been established between transition-path theory and stochastic thermodynamics suggests a novel method for rigorously grounded inference of reaction coordinates: generate an ensemble of transition paths using transition-path sampling [24, 25] or related algorithms [26–28]; estimate entropy production along chosen coordinates [29–31] or identify linear combinations of coordinates producing the most entropy using dissipative-components analysis [32]; use these most dissipative coordinates to enhance sampling of transition paths; and through further iteration identify system coordinates producing the most entropy in the transition-path ensemble and hence of most relevance to the reaction.

Machine-learning approaches to solve for high-dimensional committor coordinates [33–35] or find low-dimensional reaction models that retain predictive power [36, 37] are active areas of research [38]. The information-theoretic and thermodynamic perspectives on reactive trajectories described in this chapter provide guidance to the development of data-intensive automated methods to infer these models and their corresponding reaction mechanisms.

References

1. Zwanzig, R.: Nonequilibrium Statistical Mechanics. Oxford University Press, New York (2001)
2. Metzner, P., Schutte, C., Vanden-Eijnden, E.: Transition path theory for Markov jump processes. Multiscale Model. Simul. **7**(3), 1192–1219 (2009)
3. Vanden-Eijnden, E.: Transition path theory. In: Bowman, G.R., Pande, V.S., Noe, F. (eds.) An Introduction to Markov State Models and Their Application to Long Timescale Molecular Simulation, chapter 7, pp. 91–100. Springer, Berlin (2014)
4. Berezhkovskii, A.M., Szabo, A.: Committors, first-passage times, fluxes, Markov states, milestones, and all that. J. Chem. Phys. **150**, 54106 (2019)
5. Weinan, E., Vanden-Eijnden, E.: Towards a theory of transition paths. J. Stat. Phys. **123**(3), 503–523 (2006)
6. Cover, T.M., Thomas, J.A.: Elements of Information Theory, 2nd edn. Wiley, Hoboken (2006)
7. Busiello, D.M., Gupta, D., Maritan, A.: Entropy production in systems with unidirectional transitions. Phys. Rev. Res. **2**(2), 1–15 (2020)
8. Horowitz, J.M., Esposito, M.: Thermodynamics with continuous information flow. Phys. Rev. X **4**, 031015 (2014)
9. Esposito, M.: Stochastic thermodynamics under coarse-graining. Phys. Rev. E **85**, 041125 (2012)
10. Amari, S.-I.: Information Geometry and Its Applications, 1st edn. Springer, Tokyo (2016)
11. Nielsen, F.: An elementary introduction to information geometry. Entropy **22**, 1100 (2020)
12. Ito, S.: Stochastic thermodynamic interpretation of information geometry. Phys. Rev. Lett. **121**, 30605 (2018)
13. Ruppeiner, G.: Thermodynamics: a Riemannian geometric model. Phys. Rev. A **20**(4), 1608–1613 (1979)
14. Crooks, G.E.: Measuring thermodynamic length. Phys. Rev. Lett. **99**, 100602 (2007)
15. Tsai, S.-T., Tiwary, P.: On the distance between A and B in molecular configuration space. Mol. Simul. **47**, 449–456 (2021)
16. Hartich, D., Barato, A., Seifert, U.: Stochastic thermodynamics of bipartite systems: transfer entropy inequalities and a Maxwell's demon interpretation. J. Stat. Mech Theory Exp. **2014**, 02016 (2014)
17. Barato, A.C., Hartich, D., Seifert, U.: Rate of mutual information between coarse-grained non-Markovian variables. J. Stat. Phys. **153**, 460–478 (2013)
18. Chetrite, R., Rosinberg, M.L., Sagawa, T., Tarjus, G.: Information thermodynamics for interacting systems without bipartite structure. J. Stat. Mech. **21**, 114002 (2019)
19. Metropolis, N., Rosenbluth, A.W., Rosenbluth, M.N., Teller, A.H., Teller, E.: Equation of state calculations by fast computing machines. J. Chem. Phys. **21**, 1087 (1953)
20. Li, W., Ma, A.: Reaction mechanism and reaction coordinates from the viewpoint of energy flow. J. Chem. Phys. **144**, 114103 (2016)
21. Bolhuis, P.G., Dellago, C.: Practical and conceptual path sampling issues. Eur. Phys. J. Spec. Top. **224**, 2409–2427 (2015)
22. Peters, B.: Reaction coordinates and mechanistic hypothesis tests. Annu. Rev. Phys. Chem. **67**, 669–690 (2016)
23. Johnson, M.E., Hummer, G.: Characterization of a dynamic string method for the construction of transition pathways in molecular reactions. J. Phys. Chem. B **116**(29), 8573–8583 (2012)
24. Dellago, C., Bolhuis, P.G., Csajka, F.S., Chandler, D.: Transition path sampling and the calculation of rate constants. J. Chem. Phys. **108**, 1964 (1998)
25. Bolhuis, P.G., Chandler, D., Dellago, C., Geissler, P.L.: Transition path sampling: throwing ropes over rough mountain passes, in the dark. Annu. Rev. Phys. Chem. **53**, 291–318 (2002)
26. Van Erp, T.S., Moroni, D., Bolhuis, P.G.: A novel path sampling method for the calculation of rate constants. J. Chem. Phys. **118**, 6617 (2003)

27. Allen, R.J., Warren, P.B., Ten Wolde, P.R.: Sampling rare switching events in biochemical networks. Phys. Rev. Lett. **94**, 018104 (2005)
28. Faradjian, A.K., Elber, R.: Computing time scales from reaction coordinates by milestoning. J. Chem. Phys. **120**, 10880–10889 (2004)
29. Seifert, U.: From stochastic thermodynamics to thermodynamic inference. Annu. Rev. Condens. Matter Phys. **10**, 171–192 (2019)
30. Li, J., Horowitz, J.M., Gingrich, T.R., Fakhri, N.: Quantifying dissipation using fluctuating currents. Nat. Commun. **10**, 1666 (2019)
31. Skinner, D.J., Dunkel, J.: Improved bounds on entropy production in living systems. PNAS **118**(18), e2024300118 (2021)
32. Gnesotto, F.S., Gradziuk, G., Ronceray, P., Broedersz, C.P.: Learning the non-equilibrium dynamics of Brownian movies. Nat. Commun. **11**, 5378 (2020)
33. Khoo, Y., Lu, J., Ying, L.: Solving for high-dimensional committor functions using artificial neural networks. Res. Math. Sci. **6**, 1 (2019)
34. Li, Q., Lin, B., Ren, W.: Computing committor functions for the study of rare events using deep learning. J. Chem. Phys. **151**, 54112 (2019)
35. Rotskoff, G.M., Mitchell, A.R., Vanden-Eijnden, E.: Active importance sampling for variational objectives dominated by rare events: consequences for optimization and generalization. In: Mathematical and Scientific Machine Learning PMLR, pp. 757–780 (2022)
36. Ma, A., Dinner, A.R.: Automatic method for identifying reaction coordinates in complex systems. J. Phys. Chem. B **109**, 6769–6779 (2005)
37. Wang, Y., Tiwary, P.: State predictive information bottleneck. J. Chem. Phys. **154**, 134111 (2021)
38. Wang, Y., Ribeiro, J.M.L., Tiwary, P.: Machine learning approaches for analyzing and enhancing molecular dynamics simulations. Curr. Opin. Struct. Biol. **61**, 139–145 (2020)

Chapter 4
Multidimensional Minimum-Work Protocols

Abstract In this chapter, control protocols are designed that invert the magnetization of a 3×3 Ising model with minimal energetic cost in the slow-driving limit and compared with naive protocols. Using multiple control parameters, which provide additional flexibility in manipulating the system conformation relative to a single control parameter, allows the system to be driven along a fast-relaxing pathway between reaction endpoints. An accounting of the work done by each control parameter is also presented and analyzed (Material in this chapter also appears in Ref. Louwerse and Sivak, J. Chem. Phys. 156:194108, 2022).

4.1 Introduction

In the previous chapter, a thermodynamic perspective on reaction dynamics that allows quantification of reaction information was discussed. This provides a concrete theoretical framework for determining small sets of collective variables that are relevant to a reaction. As discussed in Sect. 2.3.1, there is interest in understanding the correspondence between relevant collective variables and efficient driving protocols. However, there has been little work in the literature to date discussing minimum-work protocols that drive multiple spatial coordinates of a system between metastable states. In this chapter, protocols in a multidimensional control-parameter space are examined to establish some intuition about principles of minimum-work driving.

Many systems of interest have slow relaxations between metastable mesostates, and finding good control parameters to drive the system can be non-trivial [1]. For finite-duration protocols where the system has insufficient time to fully relax, the system can get stuck in metastable mesostates, causing the nonequilibrium system distribution to lag the equilibrium distribution, resulting in excess work. A longer-duration protocol would allow further time for the system to relax from these metastable mesostates towards the equilibrium distribution and reduce excess work, but it may not be feasible or desirable to increase duration.

A similar problem arises in enhanced-sampling contexts, where external control parameters are used to bias system degrees of freedom that are relevant to

M. Louwerse, *Efficient Control and Spontaneous Transitions*, Springer Theses,
https://doi.org/10.1007/978-3-031-40534-1_4

characterizing metastable mesostates, reducing the time required to sample the system space and compute free energies [2, 3]. A collective variable is sought that captures the slowest relaxation mode, otherwise the system can get stuck in a metastable region, resulting in slow convergence of relevant statistics (e.g., free-energy profiles) [1, 4]. Using multiple biasing coordinates that couple to relevant relaxation modes can significantly improve sampling speed by allowing the system to more easily circumvent dynamical barriers [5–9].

Inspired by these results, multidimensional minimum-work protocols are investigated in this chapter in more detail, using external magnetic fields to drive a 3×3 Ising system between energetically stable configurations. Reducing the size of the system and control-parameter space greatly simplifies the calculations, since analytical calculations of the Boltzmann distribution and static equilibrium properties throughout control-parameter space are possible. Of particular interest is the development of a physically intuitive picture of the geometric aspects of control protocols and the generalized friction matrix. The Ising model is introduced in Sect. 4.2, then the calculations of minimum-work protocols are outlined in Sect. 4.3. Section 4.4 presents the results of the calculations and further analysis on the differences between naive and designed protocols. Section 4.5 concludes with a broader discussion of the results and their overall relevance to this thesis.

4.2 Ising System

The Ising model is a workhorse model used throughout statistical mechanics to study a variety of systems and phenomena, such as phase transitions, nucleation, and gene-regulatory networks [10]. Its basic components are a set of up or down spins (or more generally, a Bernoulli random variable) with pairwise coupling energy J. If $J > 0$, spin pairs have lower energy when they are aligned, and the Ising model is called ferromagnetic. Additionally, a magnetic field $\boldsymbol{\lambda}$ can be applied to spins to affect the relative energies of the up and down orientation of each spin.

Calculations of minimum-work protocols to drive a large 2D Ising model between the all-down and all-up configurations using multiple control parameters have been done in Refs. [11–13]. In particular, Ref. [12] drove the Ising model using a set of magnetic fields applied to blocks of spins in the 2D array, which pushes on collective variables given by the magnetization of spins under the control of a field. The minimum-work protocols designed for this control-parameter space resemble the minimum-free energy path (Sect. 2.2) calculated for this system [14].

Our system is a 2-dimensional 3×3 Ising model with fixed anti-symmetric boundary conditions [12], illustrated in Fig. 4.1. The spins are ferromagnetically coupled, with spin configuration $\{\boldsymbol{\sigma}\}$ having internal energy

$$E_{\mathrm{int}}(\{\boldsymbol{\sigma}\}) \equiv -J \sum_{\{i,j\}} \sigma_i \sigma_j \ , \tag{4.1}$$

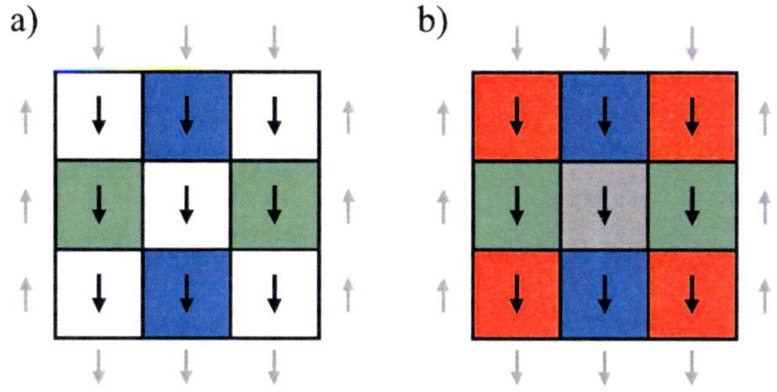

Fig. 4.1 Schematic of 3×3 Ising model with 9 fluctuating spins (black, shown in initial all-down configuration) and 12 fixed boundary spins (gray). (**a**) 2D control parameter given by two external fields on spins colored blue and green, respectively. (**b**) 4D control parameter given by four external fields on spins colored according to their symmetry type. Spins with the same color are influenced by a common magnetic field

where $J = 1\,k_\mathrm{B}T$ is the coupling coefficient for Boltzmann constant k_B and temperature T, $\sigma_i \in \{-1, 1\}$ is the orientation of spin i, and $\sum_{\{i,j\}}$ denotes a sum over nearest-neighbor spin pairs. The system evolves under single-spin-flip Glauber dynamics [15], where at the end of a fixed time interval a spin is chosen at random and a spin-flip is attempted, with acceptance probability given by the Glauber criterion (Eq. 2.6). The system has two energetically metastable configurations, with spins all down or all up, and the dynamical barrier separating the metastable configurations has activation energy $\Delta E^\ddagger = 8k_\mathrm{B}T$; the spin-inversion transition from all down to all up is a relatively rare event, with mean first-passage time of $\sim 2 \times 10^3$ attempted spin flips [16].

The spin-inversion reaction to flip the system configuration from all-down to all-up is driven by a set of external magnetic fields $\boldsymbol{h}$ applied to sets of spins (background colors in Fig. 4.1), chosen to reflect the symmetry of the boundary conditions. The applied magnetic fields $\boldsymbol{\lambda}$ bias the total energy as

$$V(\{\boldsymbol{\sigma}\}, \boldsymbol{\lambda}) \equiv E_\mathrm{int}(\{\boldsymbol{\sigma}\}) - \boldsymbol{\lambda}^\mathrm{T} \cdot \boldsymbol{X}(\{\boldsymbol{\sigma}\}) \,, \tag{4.2}$$

where $\boldsymbol{X}(\{\boldsymbol{\sigma}\})$ is a vector of conjugate forces for configuration $\{\boldsymbol{\sigma}\}$, with force $X_i \equiv -\partial V/\partial\lambda_i$ conjugate to control parameter (external field) λ_i, equaling the total magnetization of the spin set controlled by λ_i. The system magnetization is inverted by changing the magnetic field from $\lambda_{i,\mathrm{initial}} = -0.5\,k_\mathrm{B}T$ for all i (favoring the all-down configuration) to $\lambda_{i,\mathrm{final}} = 0.5\,k_\mathrm{B}T$ (favoring the all-up configuration).

Two multidimensional control-parameter spaces are explored, associated with a 2D magnetic field applied to the blue and green spins (Fig. 4.1a, the spins that are most biased by the boundary conditions) or a 4D magnetic field applied to all spins (Fig. 4.1b, the highest dimensionality that does not break the symmetry imposed by the boundary conditions). The 2D control-parameter space is a two-dimensional manifold of the 4D control-parameter space with $\lambda_\mathrm{red} = \lambda_\mathrm{black} = 0\,k_\mathrm{B}T$.

4.3 Methods

In both 2D and 4D control-parameter spaces three types of protocols between the same endpoints are considered. A naive protocol changes all magnetic fields with constant velocity. The time-optimized protocol changes all magnetic fields together (i.e., $\lambda_i(t) = \lambda_j(t)$ for all i, j, t), with velocity optimized to minimize (within the linear-response approximation) the work (see Sect. 2.3.5). Both these protocol types are equivalent to one-dimensional control using a single magnetic field to drive the system, with scalar conjugate force equal to the total magnetization of controlled spins (only blue and green spins' magnetization for 2D, and total magnetization for 4D). This yields a one-dimensional friction coefficient, the sum of all elements of the 2D or 4D friction matrices $\zeta^{1D}(\boldsymbol{\lambda}) = \sum_{ij} \zeta_{ij}(\boldsymbol{\lambda})$. During the time-optimized protocols, all fields are changed with velocity proportional to the inverse square-root of the one-dimensional friction coefficient [17], $\dot{\lambda}^{1D} \propto \zeta^{1D}(\boldsymbol{\lambda})^{-1/2}$. The fully optimized protocols solve the Euler-Lagrange equation [18] for the excess work (2.64) in all control-parameter dimensions, which is done numerically using the string method [12] (see Sect. 2.3.5).

The 4×4 friction matrix (2.62) is estimated on a discrete grid in 4D control-parameter space, then used to design fully optimized 2D and 4D protocols. Sampling the friction in a multidimensional space has a significant computational cost since the number of grid points scales exponentially with the number of control parameters; in contrast, the time-optimized protocol can be calculated by estimating the friction matrix along the one-dimensional protocol path, which has a significantly smaller computational cost. (The naive protocol requires no prior sampling.) Each protocol type is simulated for a range of protocol durations to collect an ensemble of system responses. Appendix B provides details on numerical methods.

4.3.1 Reference Relaxation Time

Throughout, work and power are reported divided by the number N_i of spins controlled by field i ($N_{\text{red}} = 4$, $N_{\text{blue}} = N_{\text{green}} = 2$, and $N_{\text{black}} = 1$). The protocol duration is scaled by a reference relaxation time τ_{rel} discussed below.

Since the system is driven from all-down to all-up, a reference relaxation time is chosen to reflect the timescale of this transition. The rate constant $k(\boldsymbol{\lambda})$ for the reaction from all-down to all-up at equilibrium for control-parameter $\boldsymbol{\lambda}$ is calculated from transition-path theory [16] (Sect. 2.2). This gives the mean number of all-down to all-up transitions observed per unit time, given that the system is initially in the all-down configuration. This rate constant is averaged over control-parameter values along the 4D naive protocol to obtain an average rate constant $\bar{k}$ for the spin inversion during this protocol:

$$\bar{k} = \frac{1}{\Delta t} \int_0^{\Delta t} \mathrm{d}t \; k(\boldsymbol{\lambda}(t)) \, . \tag{4.3}$$

$\bar{k}$ represents the mean number of transitions from all-down to all-up during a naive 4D protocol of duration Δt, assuming the system is in local equilibrium in the initial metastable basin throughout the protocol. $\tau_{\rm rel} = \bar{k}^{-1}$ then is the protocol time required for the system to make on average one spin-inversion transition. For this system, $\tau_{\rm rel} = 1102\,\mathrm{d}t$, where during each $\mathrm{d}t$ one spin flip is attempted. The moderate-duration protocols shown in Figs. 4.5, 4.8, 4.9, and 4.11 have duration $\Delta t = 1000\,\mathrm{d}t \approx 0.91\,\tau_{\rm rel}$.

4.4 Results

4.4.1 Avoiding High Friction

To illustrate general features of the multidimensional friction matrix for this system, Fig. 4.2a shows the trace (a scalar capturing the essential features) of the 2×2 friction matrix in the control-parameter space defined by axes $(\lambda_{\rm blue}, \lambda_{\rm green})$. The friction is maximized at $(1, -1)$, where the fields cancel the fixed boundary conditions (Fig. 4.1), stabilizing the all-down and all-up configurations relative to the unperturbed system ($\lambda_i = 0$ for all i). Both components of the friction (2.67), the force covariance (Fig. 4.2b) and the integral relaxation time (Fig. 4.2c), are similarly peaked.

Figure 4.2 also shows the 2D time-optimized and fully optimized protocols. (The naive protocol follows the same path as the time-optimized protocol but with constant protocol velocity.) The naive and time-optimized protocols follow a straight line between the fixed endpoints and drive the system through a high-friction region, with the time-optimized protocol slowing down where the friction along the path is large, near control parameters corresponding to the unperturbed system. In contrast, the fully optimized protocol avoids this by making an excursion through low-friction regions of control-parameter space.

Figure 4.3 shows the three protocol types in 2D and 4D control-parameter spaces. The time-optimized protocols have high initial and final velocity, and low velocity in the middle of the protocol, where fields are close to zero and friction is large. This is consistent with the minimum-work protocol for one-dimensional barrier crossing [19], where the protocol slows to allow time for stochastic fluctuations to kick the system over an energy barrier, thereby remaining closer to equilibrium relative to the naive protocol.

The fully (both temporally and spatially) optimized protocols depart dramatically from the naive and time-optimized protocols. Consistent with results in a larger Ising

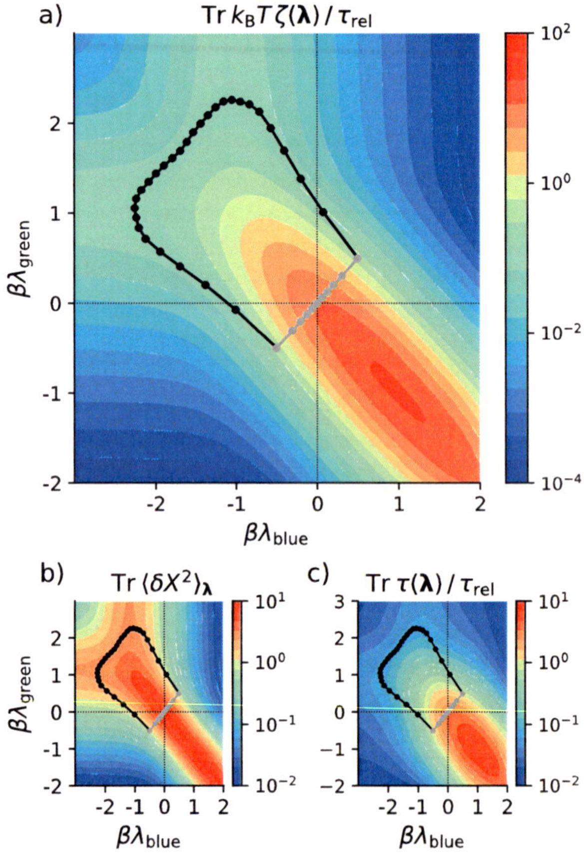

Fig. 4.2 2D fully optimized (black) and time-optimized (gray) protocols overlaid on a heatmap showing the trace of the (**a**) friction (2.62), (**b**) force covariance, or (**c**) integral relaxation time (2.68) matrix. Points denote control-parameter values evenly spaced in time on the given protocol

model [12], the green field (whose corresponding spins are initially anti-aligned with their boundary conditions and therefore energetically frustrated) is increased rapidly in the early stages of the protocol, while the blue field (whose corresponding spins are initially aligned with their boundary conditions) is decreased initially, then increased later. 2D and 4D protocols show the same trend, but the addition of time-varying red and black fields in 4D protocols shifts the relative timing of changes in green and blue fields.

A striking feature of both the 2D and 4D fully optimized protocols is the non-monotonic magnetic fields at early and late stages. In the linear-response regime, a 1D minimum-work protocol must be monotonic since a non-monotonic protocol drives through the same control-parameter value more than once, producing unnecessary dissipation; in the metric language of the linear-response approximation, the shortest curve between two points cannot cross itself. But individual fields for minimum-work protocols can be non-monotonic as long as the protocol path in multidimensional space does not loop back on itself.

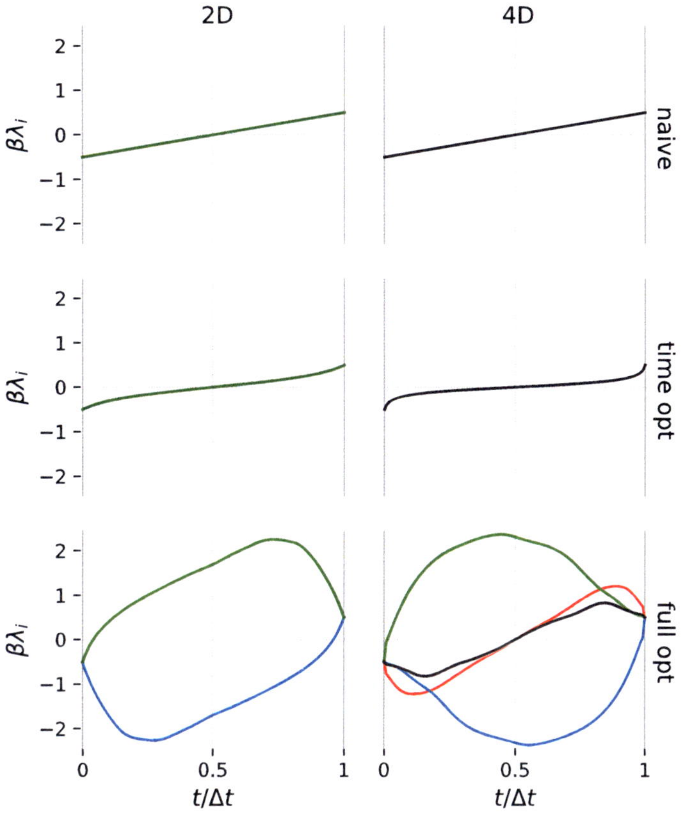

Fig. 4.3 Designed protocols for driving spin-inversion in the 3 × 3 Ising model. Magnetic fields λ_i as a function of scaled protocol time $t/\Delta t$ during naive (top row), time-optimized (middle), and fully optimized (bottom) protocols in 2D (left column) and 4D (right) control-parameter spaces. Colors correspond to spin sets in Fig. 4.1

4.4.2 Reducing Excess Work and Keeping Close to Equilibrium

Next, the excess work for all protocols is compared to assess their relative performance. Figure 4.4a shows the excess work as a function of protocol duration. In the limit of long duration, the excess works approach their respective linear-response approximations and scale as Δt^{-1}. For short protocols, the excess works approach that of an instantaneous protocol, given by the relative entropy of the initial and final equilibrium distributions [20]. Figure 4.4b shows the ratio of excess work for naive relative to time-optimized and fully optimized protocols. While the 2D and 4D time-optimized protocols respectively do $\sim 1.2\times$ and $\sim 2\times$ less excess work than the corresponding naive protocols, the fully optimized protocols respectively do $\sim 9\times$ and $\sim 26\times$ less excess work than the naive protocols. (Fully optimized

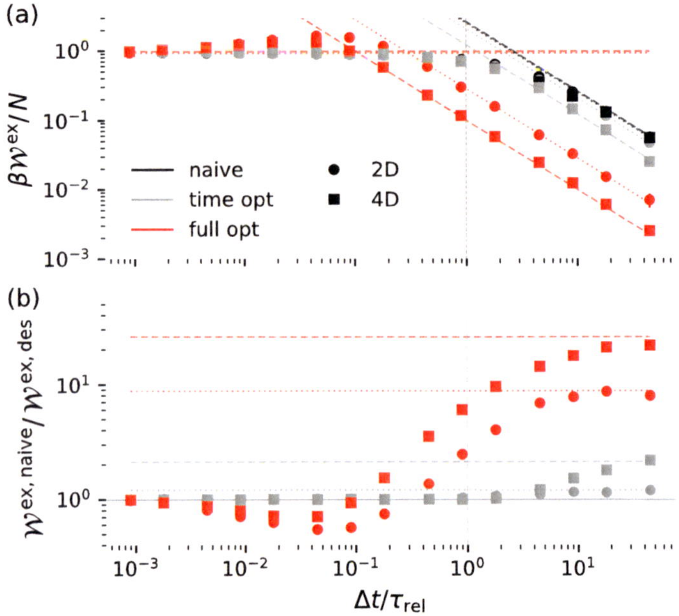

Fig. 4.4 Comparison of excess work in naive and designed protocols. (**a**) Excess work (2.64) scaled by number N of controlled spins (i.e., $N = 4$ controlled spins for 2D protocols and $N = 9$ for 4D) as a function of protocol duration Δt, for each of the six protocol types (Fig. 4.3): 2D (circles) and 4D (squares) control-parameter spaces; and naive (black), time-optimized (gray), and fully optimized (red). (**b**) Ratio of excess work during naive protocols to excess work during time-optimized (gray) and fully optimized (red) protocols for 2D (circles) and 4D (squares) control. Horizontal dotted lines: excess work for an instantaneous protocol. Dashed lines: linear-response approximations, accurate at long duration. Protocol durations are scaled by reference relaxation time τ_{rel}

control protocols do more work than all other protocols (including an instantaneous protocol) for short durations ($\Delta t/\tau_{rel} \lesssim 0.1$).)

Fully optimized protocols keep the system much closer to equilibrium than naive and time-optimized protocols for a fixed duration, resulting in significant work reduction. Figure 4.5 shows for moderate protocol duration ($\Delta t \approx \tau_{rel}$) the quasistatic and nonequilibrium mean magnetizations $\langle\sigma_i\rangle$ of each spin type. In fully optimized protocols, there is a rough ordering of spin flips matching the corresponding field increases (Fig. 4.3): green spins flip first, red and black spins flip next, and blue spins flip last (also depicted schematically in Fig. 4.6). This temporal separation of spin flips contrasts with naive and time-optimized protocols, where the different spin types have nearly equal mean magnetization throughout the protocol.

Throughout the naive and time-optimized protocols, the system remains bistable, with the external fields primarily biasing the relative energies of the all-down and all-up configurations. For the moderate duration shown, the system gets stuck in the initial metastable basin so nonequilibrium mean magnetizations significantly

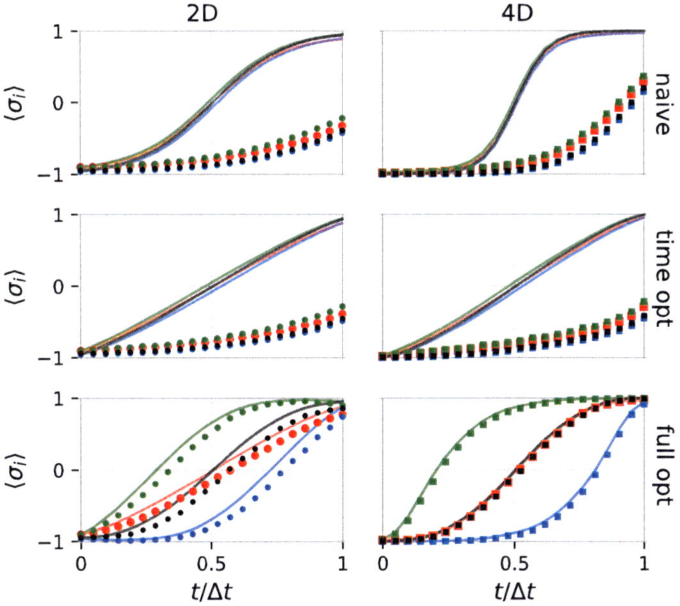

Fig. 4.5 Quasistatic (curves) and nonequilibrium (points) mean magnetization of each spin type (different colors, see Fig. 4.1) as a function of elapsed time t along a moderate-duration ($\Delta t \approx \tau_{\text{ref}}$) protocol of each type (see Fig. 4.3)

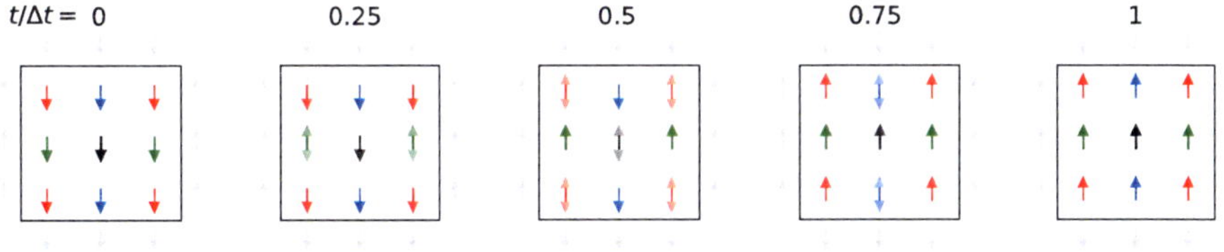

Fig. 4.6 Spin-inversion mechanism at five evenly spaced times during the 4D fully optimized protocol. The relative equilibrium population of up and down orientations are indicated by the transparency of the arrowheads

lag the quasistatic limit. In contrast, the fully optimized protocols keep the system much closer to equilibrium, with modest lag during the 2D protocol and nearly no lag during the 4D protocol.

4.4.3 Reducing Spin Covariance and Flattening the Energy Landscape

Since the fully optimized protocols drive the system through configuration space differently than the naive and time-optimized protocols, it is helpful to understand

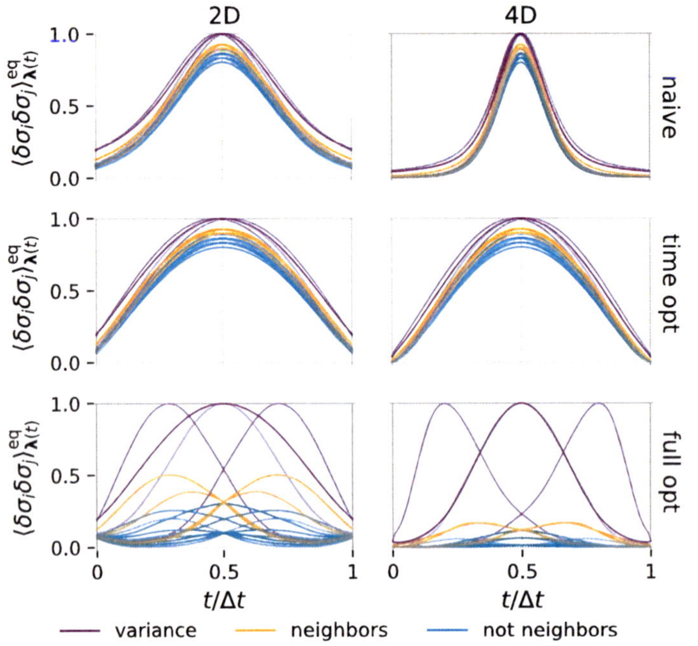

Fig. 4.7 Equilibrium spin covariance as a function of elapsed time t along a protocol of each type (Fig. 4.3). Purple: spin variance. Gold: covariance between neighboring spins. Blue: covariance between non-neighboring spins

general features that are associated with reduced protocol work in the fully optimized protocols.

Figure 4.7 shows the equilibrium variance of each spin and covariance between neighboring and non-neighboring spins. The 9×9 spin-covariance matrix is "coarse-grained" (by summing spin-covariance elements for the same spin types into a single entry) to form the 4×4 force-covariance matrix, a factor in the friction (2.67) that contributes to excess power for 4D protocols. Since each spin magnetization is a Bernoulli random variable, it is a quadratic function of the mean, $\langle \delta\sigma_i^2 \rangle = 1 - \langle \sigma_i \rangle^2$, where each variance peaks when its mean changes sign (Fig. 4.5). At some point during a long-duration protocol, the mean magnetization of each spin must change sign and hence the corresponding diagonal element of the force-covariance matrix must reach a fixed maximal value; however, the contribution from off-diagonal elements (covariance between distinct spins) can be reduced. In the naive and time-optimized protocols, where the system primarily occupies the all-up and all-down configurations, all spin covariances (even for non-neighboring spin pairs) are large where those configurations are equally populated. The fully optimized protocols disrupt this bistability and lower the total energy of many other configurations; this reduces spin covariance, with low covariance between neighboring spins and near-zero covariance between non-neighboring spins. The fully optimized 4D protocol, with additional controls, reduces covariance more than the 2D protocol.

The covariance between spins reflects the internal energy that couples their orientations; high covariance stems from ensembles dominated by low-internal-energy configurations with aligned spins, while low covariance results from the system accessing configurations with anti-aligned spins and hence higher internal energy. Next, the energetic and entropic properties of instantaneous system distributions along the control protocol are calculated.

Figure 4.8 shows the free energy along each protocol, as well as its component energies (4.2) and entropy $H \equiv -\sum_{\{\sigma\}} p(\{\boldsymbol{\sigma}\}) \ln p(\{\boldsymbol{\sigma}\})$ for (equilibrium or nonequilibrium) probability distribution $p(\{\boldsymbol{\sigma}\})$. The fully optimized protocols drive the system through distributions with high mean internal energy (4.1). This indicates significant population of high-internal-energy configurations (i.e., with anti-aligned

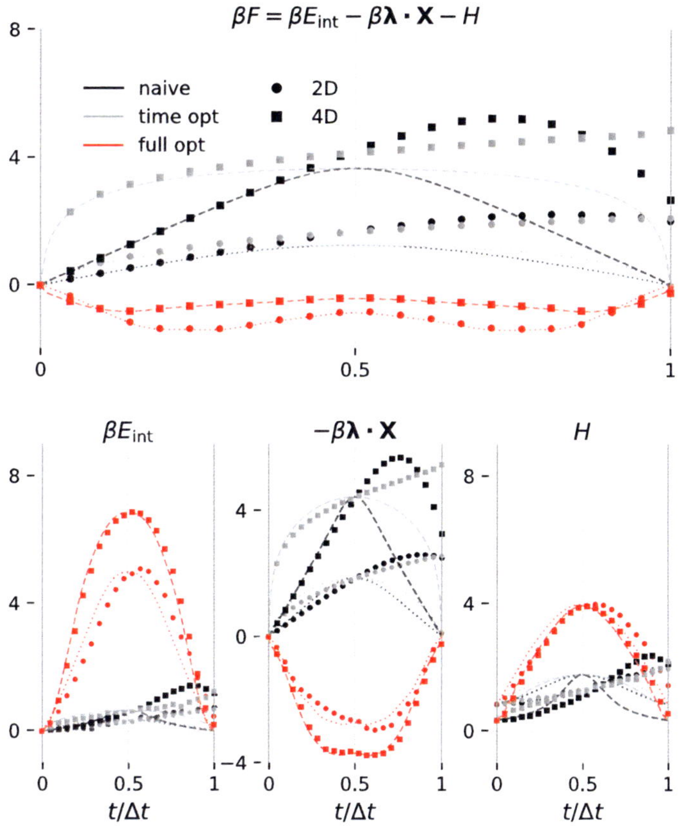

Fig. 4.8 Free energy (top), and its components the mean internal energy (4.1) (bottom left), mean external energy (bottom middle), and entropy (bottom right), of equilibrium (curves) and nonequilibrium (points) distributions, as a function of elapsed time t along a moderate-duration ($\Delta t \approx \tau_{\text{rel}}$) protocol of each type

spins), relative to naive and time-optimized protocols where the internal energy is relatively constant (consistent with the system primarily occupying the all-down or all-up configuration). These configurations with high internal energy are stabilized by stronger external fields, flattening the total-energy landscape relative to naive and time-optimized protocols. The fully optimized protocols also significantly increase the entropy, indicating many accessible system configurations, in contrast to naive and time-optimized protocols which primarily fluctuate near the all-up and all-down configurations.

The flattened total-energy landscape and increased entropy during the fully optimized protocol results in relatively constant free energy, requiring comparatively low-magnitude quasistatic work throughout the protocol. In contrast, the naive and time-optimized protocols drive the system through a substantial free-energy barrier, requiring quasistatic work input to the system during the first half of the protocol and quasistatic work extraction during the second half.

4.4.4 Heterogeneously Inputting and Extracting Work, Coinciding with System Relaxation

I now investigate how fully optimized protocols drive each spin set differently than naive and time-optimized protocols, by splitting total work into quasistatic and excess works by each control-parameter component:

$$\mathcal{W} = \sum_i \left[\mathcal{W}_i^{\text{qs}} + \mathcal{W}_i^{\text{ex}}\right] \tag{4.4a}$$

$$= \sum_i \left[\int_0^{\Delta t} \mathrm{d}t\, \mathcal{P}_i^{\text{qs}}(t) + \int_0^{\Delta t} \mathrm{d}t\, \mathcal{P}_i^{\text{ex}}(t)\right] \tag{4.4b}$$

$$\approx \sum_i \left[-\int_0^{\Delta t} \mathrm{d}t\, \dot{\lambda}_i(t)\langle X_i\rangle_{\boldsymbol{\lambda}(t)}^{\text{eq}} + \int_0^{\Delta t} \mathrm{d}t \sum_j \dot{\lambda}_i(t)\zeta_{ij}(\boldsymbol{\lambda}(t))\dot{\lambda}_j(t)\right]. \tag{4.4c}$$

The final line uses the linear-response approximation (2.64). The quasistatic works of all components sum to the path-independent equilibrium free-energy difference between control-protocol endpoints, $\sum_i \mathcal{W}_i^{\text{qs}} = \Delta F$; however, the quasistatic work of each component is path-dependent.

Figure 4.9 shows components of the quasistatic power $\mathcal{P}_i^{\text{qs}}(t)$ and finite-duration power $\mathcal{P}_i(t)$, for moderate protocol duration ($\Delta t \approx \tau_{\text{rel}}$). During naive and time-optimized protocols, each field inputs and extracts work equally, inputting work in roughly the first half of the protocol to increase the system's free energy and extracting it in the second half to reduce the free energy (Fig. 4.8). For finite duration, the mean conjugate force lags its equilibrium mean (Fig. 4.5), and the quasistatic work is not fully extracted in the second half of the protocol, resulting

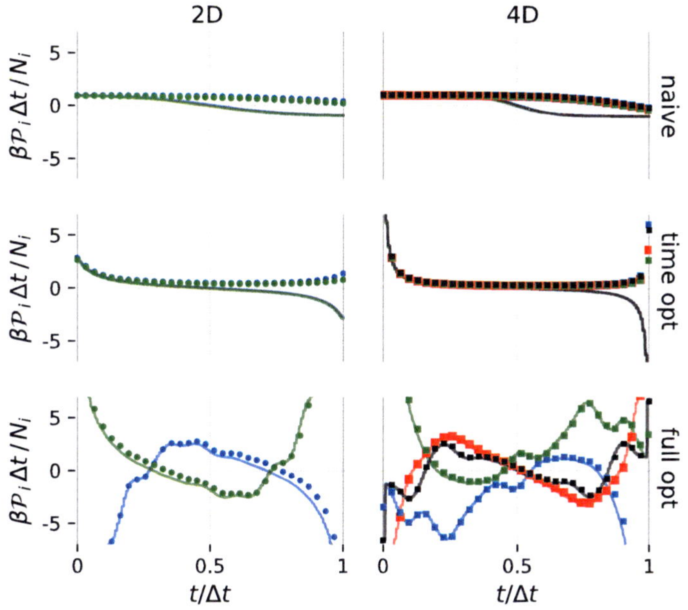

Fig. 4.9 Quasistatic power (curves) and finite-duration power (Eq. (4.4b), points) of each field (colors, see Fig. 4.1) as a function of elapsed time t along a moderate-duration ($\Delta t \approx \tau_{\rm rel}$) protocol of each type (see Fig. 4.3). Power $\mathcal{P}_i$ is scaled by the number N_i of spins controlled by field i

in excess work. The fully optimized protocols input and extract work differently through the different fields, with zero power for each component where either the field velocity is zero or the spin magnetization for that field is zero (i.e., controlled spins on average flip from down to up).

The 2D fully optimized protocol is roughly divided into three stages according to sign changes of the power. Until $t/\Delta t \approx 0.3$, the green field is increased while the blue field is decreased (Fig. 4.3), inputting power to the green spins to increase their mean magnetization (Fig. 4.5) while power is extracted from the blue spins (which do not significantly change mean magnetization), so that overall nearly no power is input to the system and the free energy is relatively constant (Fig. 4.8). Second (until $t/\Delta t \approx 0.7$), both fields are increased and work is input to the blue spins to increase their magnetization, while work is extracted from the green spins as they are stabilized in spin-up. Finally, the green field is reduced and work is input to the green spins, while increasing the blue field stabilizes the spin-up configuration of the blue spins and extracts work: the two fields act in concert to keep the free energy more constant. This elaborate and heterogeneous schedule of work input and extraction could reflect the system's energetic requirements to fluctuate into configurations with high internal energy: during naive and time-optimized protocols the system must wait for appropriate fluctuations from the environment to overcome the internal energy barrier separating the all-down and all-up configurations in order to relax to equilibrium; in contrast, the fully optimized protocols externally provide

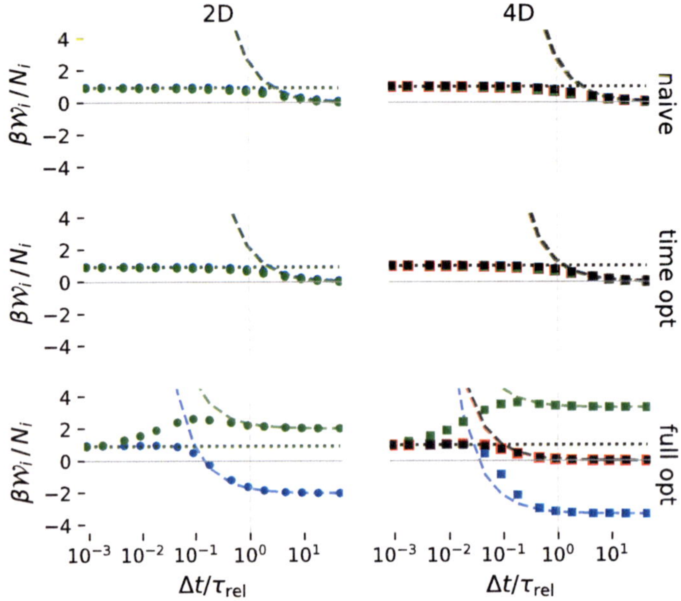

Fig. 4.10 Work (4.4a) as a function of protocol duration Δt for each field (colors, see Fig. 4.1), for each protocol type (see Fig. 4.3). Dotted lines: instantaneous work. Dashed curves: linear-response approximation. Protocol durations are scaled by reference spin-inversion relaxation time τ_{rel}

the appropriate energy to each spin as work, allowing the system to access high-internal-energy configurations and significantly decreasing the time required for system relaxation. The 4D fully optimized protocol shows similar features to 2D in the green and blue power, but events happen relatively earlier in green and later in blue.

Figure 4.10 shows the work $\mathcal{W}_i$ by each field i as a function of protocol duration, asymptoting to the instantaneous work $(-[\lambda_{i,\text{final}} - \lambda_{i,\text{initial}}]\langle X_i \rangle^{\text{eq}}_{\boldsymbol{\lambda}(t=0)} \approx 1\, k_{\text{B}}T$ per spin) for short durations and reaching the (path-dependent) quasistatic work in the long-duration limit. For the naive and time-optimized protocols, each field does the instantaneous work until $\Delta t \approx \tau_{\text{rel}}$ before crossing over to near-zero quasistatic values for longer durations.

The fully optimized protocols input work via the green field and extract work via the blue field in the quasistatic limit. Work input from the green field increases from instantaneous to quasistatic values at short durations while work inputs from other fields remain near instantaneous values. When the duration is sufficiently long for the system to relax ($\Delta t \approx 0.1\tau_{\text{rel}}$), the work from the remaining fields crosses over from instantaneous to quasistatic values, with the blue field extracting work from the system for moderate-to-long duration. This suggests that work input by the green field early in the protocol is transduced through the system to be extracted by the blue field later in the protocol. Driving the green spins to flip early in the protocol initiates the transition mechanism, with further work input to red and black spins continuing the complete spin inversion by allowing the system to access high-internal-energy configurations, and final work extraction from blue spins stabilizing the system in the low-internal-energy all-up configuration. Providing work to specifically support this transition mechanism—rather than simply homogeneously inputting and extracting work to each spin type—keeps the system close to equilibrium for durations $\sim 10\times$ shorter than naive and time-optimized protocols.

Figure 4.11 shows the excess power by each field for moderate duration ($\Delta t \approx \tau_{\mathrm{rel}}$), as well as linear-response approximations (4.4b). The excess power during naive and time-optimized protocols is essentially equal for each field, consistent with the field velocities (Fig. 4.3) and lag of conjugate forces (Fig. 4.5) being equal for each component in these protocols. The linear-response approximation for the excess power during naive protocols peaks in the middle where friction is largest (Fig. 4.2). In time-optimized protocols, the linear-response approximation to the excess power is essentially the same for each field and constant, flattening the excess-power peak in naive protocols by reducing protocol velocity in this region. The excess power at this moderate duration (significantly distant from equilibrium) unsurprisingly differs from the linear-response approximation for both naive and time-optimized protocols. Nevertheless, the excess power during the fully optimized protocol is significantly reduced relative to both naive and time-optimized protocols

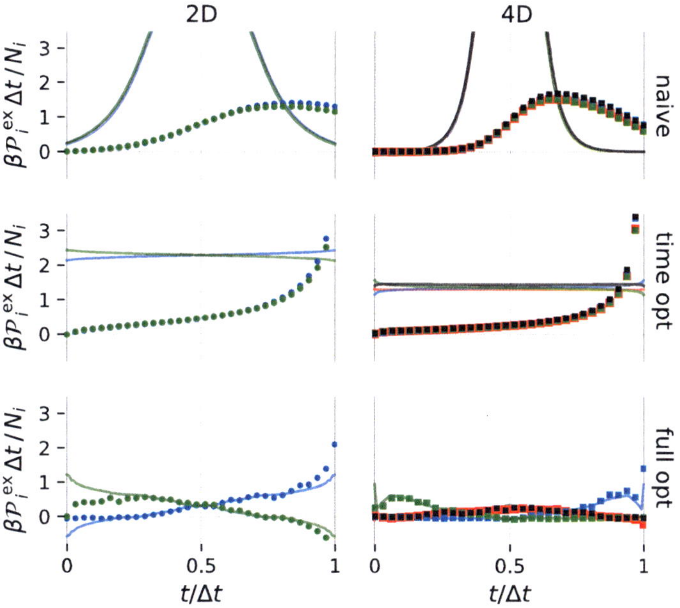

Fig. 4.11 Exact excess power (points, Eq. (4.4b)) and linear-response approximation (curves, Eq. (4.4c)) for each field (colors, see Fig. 4.1) as a function of elapsed time t along a moderate-duration ($\Delta t \approx \tau_{\mathrm{rel}}$) protocol of each type (Fig. 4.3)

and agrees well with the linear-response approximation, indicating that the system remains much closer to equilibrium.

4.5 Discussion

The naive and time-optimized protocols (which follow the same path through multidimensional control-parameter space) change all fields simultaneously, biasing the total magnetization to drive the system from all-down to all-up. The barrier separating the energetically metastable configurations remains large throughout the protocol, and for moderate-duration protocols, the slow transitions between these endpoints prevent the system from relaxing to equilibrium, resulting in significant lag and large excess work. The time-optimized protocol reduces velocity where the friction is large to allow more time for system relaxation, which for moderate-to-long durations reduces the excess work relative to the naive protocol.

In contrast, fully optimized protocols in multidimensional control-parameter spaces have more flexibility to manipulate the system's total energy (4.2); they use this flexibility to flatten the total-energy landscape through a heterogeneous schedule of work input and extraction through each spin (Figs. 4.8 and 4.9), also speeding up system transitions from all-down to all-up. Both these equilibrium thermodynamic and kinetic properties are captured by the generalized friction matrix through the respective matrices of force covariance and integral relaxation time (2.67). Both components of the friction play a role in improved understanding of minimum-work control in multidimensional control-parameter spaces.

The force covariance matrix is a property of the equilibrium distribution without any reference to the system's dynamics. It quantifies the curvature of equilibrium free energy in control-parameter space [21]. For systems where the integral relaxation time is constant, the force covariance matrix replaces the generalized friction as the metric on control-parameter space that quantifies excess work, indicating that constant free-energy paths minimize work. Although the integral relaxation time is not constant throughout control-parameter space for this system (Fig. 4.2c), fully optimized protocols still have relatively constant free energy relative to naive and time-optimized protocols (Fig. 4.8), a factor that helps to reduce work.

Off-diagonal elements of the force covariance matrix in fully optimized protocols are significantly reduced relative to naive and time-optimized protocols (Fig. 4.7), reflecting that the system accesses configurations which are otherwise inaccessible due to high internal energy. This has interesting connections to strategies for optimal inference of system parameters [22]. There, a spin system is perturbed by external fields to new equilibrium ensembles that allow more efficient inference of system coupling constants (i.e., fewer samples are required to estimate parameters to a given precision), and the perturbation is updated using the Fisher information matrix for desired system parameters. Inference is optimal at field strengths that decorrelate spins, allowing the system to occupy otherwise-inaccessible configurations. Since the excess work during the protocol is known to affect the efficiency of free-energy estimation [23–25] and the force covariance matrix is proportional to the Fisher information matrix [21], it is not surprising that parallels in design strategy exist. However, a key distinction is the dynamical nature of control protocols, where the relaxation kinetics between equilibrium distributions of successive control parameters is relevant; thus the importance of the generalized friction, which captures relaxation timescales via the integral relaxation time matrix.

The fully optimized protocols reveal a clear mechanism for spin inversion. First, work is input to green spins (initially energetically frustrated) causing them to flip, which together with direct work input by red and black fields drives neighboring red and black spins to flip. At the midpoint of the 4D fully optimized protocol, both green spins are up, red and black spins are evenly split between up and down, and blue spins are both down. Past the midpoint, work is extracted from red and black spins by increasing the corresponding fields to stabilize spins in the up orientation, which together with work input to neighboring blue spins causes blue spins to flip. Finally, work is extracted from blue spins to stabilize the system in the all-up configuration.

Using this strategy, the system reaches the linear-response regime for durations $\sim 10\times$ shorter than in the naive and time-optimized protocols (Fig. 4.10). Excess work scales as Δt^{-1} for long-duration protocols where the system is in the linear-response regime. A protocol that enables fast relaxation so the system reaches this regime and excess-work scaling at relatively short duration results in work reduction relative to a naive protocol that reaches the linear-response regime at longer duration. The fully optimized protocols appear to facilitate such fast-relaxing paths in configuration space, reducing the time spent in the initial metastable basin before making a transition over the energy barrier. Earlier studies showed an empirical similarity between the minimum-work protocol and the minimum free-energy path [12, 14], which represents the spontaneous (in the absence of driving) transition path between metastable mesostates. The minimum-work protocols here also drive the system along a plausible spontaneous transition path: green spins flip first, followed by red and black spins, and finally blue spins. There may be a deeper connection between minimum-work protocols and spontaneous transition paths of the unperturbed system, that would allow extraction of kinetic information about transition paths from minimum-work protocols or use kinetic information to design efficient protocols. A more detailed quantitative comparison is necessary to elucidate this connection.

References

1. Moradi, M., Tajkhorshid, E.: Computational recipe for efficient description of large-scale conformational changes in biomolecular systems. J. Chem. Theory Comput. **10**(7), 2866–2880 (2014)
2. Tiwary, P., van de Walle, A.: A review of enhanced sampling approaches for accelerated molecular dynamics. In: Weinberger, C.R., Tucker, G.J. (eds.) Multiscale Materials Modeling for Nanomechanics, chap. 6, pp. 195–221. Springer International Publishing, New York (2016)

3. Yang, Y.I., Shao, Q., Zhang, J., Yang, L., Gao, Y.Q.: Enhanced sampling in molecular dynamics. J. Chem. Phys. **151**(7) (2019)
4. Morfill, J., Neumann, J., Blank, K., Steinbach, U., Puchner, E.M., Gottschalk, K.E., Gaub, H.E.: Force-based analysis of multidimensional energy landscapes: application of dynamic force spectroscopy and steered molecular dynamics simulations to an antibody fragment-peptide complex. J. Mol. Biol. **381**(5), 1253–1266 (2008)
5. Maragliano, L., Vanden-Eijnden, E.: A temperature accelerated method for sampling free energy and determining reaction pathways in rare events simulations. Chem. Phys. Lett. **426**(1–3), 168–175 (2006)
6. Pfaendtner, J., Bonomi, M.: Efficient sampling of high-dimensional free- energy landscapes with parallel bias metadynamics. J. Chem. Theory Comput. **11**(11), 5062–5067 (2015)
7. Jiang, W., Luo, Y., Maragliano, L., Roux, B.: Calculation of free energy land- scape in multi-dimensions with Hamiltonian-exchange umbrella sampling on petascale supercomputer. J. Chem. Theory Comput. **8**(11), 4672–4680 (2012)
8. Zhao, T., Fu, H., Lelièvre, T., Shao, X., Chipot, C., Cai, W.: The extended generalized adaptive biasing force algorithm for multidimensional free-energy calcula- tions. J. Chem. Theory Comput. **13**(4), 1566–1576 (2017)
9. Chipot, C., Lelièvre, T.: Enhanced sampling of multidimensional free-energy landscapes using adaptive biasing forces. SIAM J. Appl. Math **81**(5), 1673–1695 (2011)
10. Plischke, M., Bergersen, B.: Equilibrium Statistical Physics, 3rd edn. World Scientific Publishing, Toh Tuck Link, Singapore (2006)
11. Rotskoff, G.M., Crooks, G.E.: Optimal control in nonequilibrium systems: dynamic Riemannian geometry of the Ising model. Phys. Rev. E **92**, 060102 (2015)
12. Rotskoff, G.M., Crooks, G.E., Vanden-Eijnden, E.: A geometric approach to optimal nonequilibrium control: minimizing dissipation in nanomagnetic spin systems. Phys. Rev. E **95**, 012148 (2017)
13. Gingrich, T., Rotskoff, G., Crooks, G., Geissler, P.: Near-optimal protocols in complex nonequilibrium transformations. PNAS **113**, 10263–10268 (2016)
14. Venturoli, M., Vanden-Eijnden, E., Ciccotti, G.: Kinetics of phase transitions in two dimensional Ising models studied with the string method. J. Math. Chem. **45**, 188–222 (2009)
15. Glauber, R.J.: Time-dependent statistics of the Ising model. J. Math. Phys. **4**(2), 294–307 (1963)
16. Vanden-Eijnden, E.: Transition path theory. In: Bowman, G.R., Pande, V.S., Noe, F. (eds.) An Introduction to Markov State Models and Their Application to Long Timescale Molecular Simulation, chap. 7, pp. 91–100. Springer, Berlin (2014)
17. Sivak, D.A., Crooks, G.E.: Thermodynamic metrics and optimal paths. Phys. Rev. Lett. **108**(19), 190602 (2012)
18. Morin, D.: Introduction to Classical Mechanics. Cambridge University Press, Cambridge (2008)
19. Sivak, D.A., Crooks, G.E.: Thermodynamic geometry of minimum-dissipation driven barrier crossing. Phys. Rev. E **94**, 052106 (2016)
20. Large, S.J., Sivak, D.A.: Optimal discrete control: minimizing dissipation in discretely driven nonequilibrium systems. J. Stat. Mech. Theory Exp. **2019**(8) (2019)
21. Crooks, G.E.: Measuring thermodynamic length. Phys. Rev. Lett. **99**, 100602 (2007)
22. Jiang, J., Sivak, D.A., Thomson, M.: Active learning of spin network models (2007). arXiv:1903.10474
23. Shenfeld, D.K., Xu, H., Eastwood, M.P., Dror, R.O., Shaw, D.E.: Mini- mizing thermodynamic length to select intermediate states for free-energy calculations and replica-exchange simulations. Phys. Rev. E **80**, 046705 (2009)
24. Kim, S., Kim, Y.W., Talkner, P., Yi, J.: Comparison of free-energy estima- tors and their dependence on dissipated work. Phys. Rev. E **86**(4), 1–10 (2012)
25. Blaber, S., Sivak, D.A.: Skewed thermodynamic geometry and optimal free energy estimation. J. Chem. Phys. **153**(24) (2020)

Chapter 5
Connections Between Minimum-Work Protocols and Transition Paths

Abstract In this chapter, a direct comparison between the designed minimum-work protocols from Chap. 4 and the spontaneous transition mechanism for magnetization inversion in a 3×3 Ising model is presented. Designed protocols are found to capture general features of the spontaneous mechanism and energetics given the constraints on the control parameters. Based on these results, the conditions under which a strong correspondence between efficient control protocols and spontaneous transition mechanisms could be found are discussed (Material in this chapter also appears in Ref. Louwerse, Thesis code for 2D Ising model, 2022. https://github.com/mdlouwerse/MDL_thesis_code).

5.1 Introduction

Here the correspondence between the system's response to a minimum-work protocol and its dynamics during a spontaneous transition is investigated. The 3×3 ferromagnetic Ising system (Fig. 5.1) is examined as it spontaneously transitions from the all-down to all-up configuration with no external magnetic fields ($\boldsymbol{\lambda} = 0$), and as it is driven between all-down and all-up configurations by a set of time-dependent magnetic fields (Chap. 4). This chapter focuses on the 4D fully optimized protocol since it drives all spins in the system and is the highest-dimension protocol calculated for this system. A sufficiently long protocol duration, $\Delta t = \tau_{\text{ref}}$, such that the system is in the linear-response regime is considered so that the system is near equilibrium (although excess work is still done on the system).

Two aspects of the reaction are of interest to this study: the reaction mechanism and its transient energetic costs. In this context, the reaction mechanism is a sequence of spin flips that brings the system from all down to all up; since there are 9 spins, there are 9 steps to the reaction mechanism. However, the system response to control protocols and its spontaneous transition paths are both ensembles of trajectories where any spin can (and frequently does) flip orientations multiple times throughout a single trajectory. Section 5.2 first considers how information about the reaction mechanism can be extracted from the trajectory ensembles.

M. Louwerse, *Efficient Control and Spontaneous Transitions*, Springer Theses,
https://doi.org/10.1007/978-3-031-40534-1_5

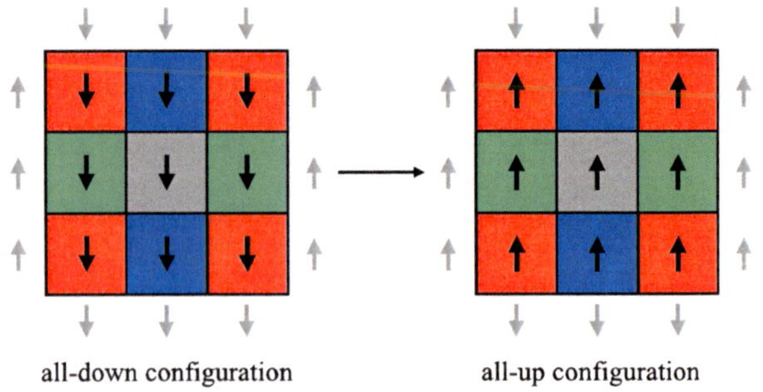

Fig. 5.1 Schematic of 3×3 Ising model with 9 fluctuating spins (black, shown in initial all-down configuration) and 12 fixed boundary spins (gray). Spins are colored according to their symmetry type

The energetic cost of the mechanism is also characterized. Control protocols do work on the system, providing an energetic bias that produces a system response. Chapter 4 shows that during the fully optimized 4D protocol, work is done by green fields early in the protocol, which flips green spins and increases the system's internal energy, and then later blue fields extract work from blue spins as they flip and decrease the system's internal energy (Fig. 4.10). Spontaneous transition paths similarly take in heat to overcome the energy barrier, but must wait for the appropriate fluctuations from the environment to complete a successful transition path. The random fluctuations that do give rise to transition paths are characterized by transition rates which are biased relative to the equilibrium detailed-balance transition rates, as described in Sect. 3.2, and therefore the heat flows associated with these conditional transition rates can be determined. Energy flows into the system can also be assigned to different degrees of freedom, which provides some insight into the reaction mechanism [1, 2]. These energetic changes in the Ising system will be examined during both the driven and spontaneous processes.

5.2 Trajectory Ensembles from Minimum-Work Protocols and the Transition-Path Ensemble

By repeating a control protocol many times, an ensemble of system responses can be generated that (roughly) transit from the all-down to all-up configurations. A sample of 5000 trajectories is generated in response to a control protocol by simulating the 4D fully optimized protocol for $\Delta t = \tau_{\text{rel}}$ as described in Appendix B.

Similarly, the transition-path ensemble (TPE) is an ensemble of reactive trajectories from all down to all up. To generate an ensemble of transition paths, the forward committor is calculated by solving the linear equation (Eq. (2.31))

$$\sum_{\boldsymbol{\phi}'} T_{\boldsymbol{\phi}'\boldsymbol{\phi}} q(\boldsymbol{\phi}') = 0 \tag{5.1}$$

with boundary conditions $q(\boldsymbol{\phi}) = 0$ for $\boldsymbol{\phi}$ the all-down configuration and $q(\boldsymbol{\phi}) = 1$ for $\boldsymbol{\phi}$ the all-up configuration. Because the forward and backward committors for an equilibrium system are related by detailed balance, the superscript in $q^{+}(\boldsymbol{\phi})$ is eliminated and the forward committor is referred to as the committor throughout this chapter.

After calculating the committor for the Ising system, trajectory segments belonging to the transition-path ensemble are generated using modified transition rates (Eq. (3.20), [3]). A trajectory is initialized in the all-down configuration $\boldsymbol{\phi}' = A$, then the next state $\boldsymbol{\phi}$ outside of A is chosen with probability

$$p^{\text{initial}}(\boldsymbol{\phi}) = \frac{T_{\boldsymbol{\phi}\boldsymbol{\phi}'} q(\boldsymbol{\phi})}{\sum_{\boldsymbol{\phi}''} T_{\boldsymbol{\phi}''\boldsymbol{\phi}'} q(\boldsymbol{\phi}'')} , \tag{5.2}$$

where the denominator normalizes this probability over all possible transitions out of A. Subsequent states are generated using modified transition rates, which for the forward TPE are

$$T^{\text{R}}_{\boldsymbol{\phi}\boldsymbol{\phi}'} = T_{\boldsymbol{\phi}\boldsymbol{\phi}'} \frac{q(\boldsymbol{\phi})}{q(\boldsymbol{\phi}')} . \tag{5.3}$$

Trajectories are terminated once they reach B for the first time. A sample of 40, 000 transition paths are generated using this method. Figure 5.2a shows a sample of these

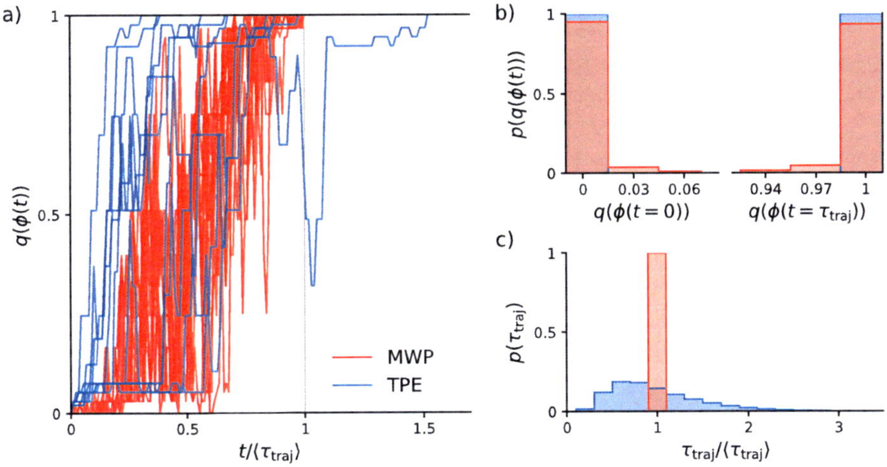

Fig. 5.2 Trajectory ensembles from the minimum-work protocol and transition-path ensemble. (**a**) 10 sample trajectories generated by an MWP (red) and from the transition-path ensemble (blue), projected onto the committor $q(\boldsymbol{\phi})$. The trajectory time is scaled by the mean duration τ_{traj} of trajectories in the respective ensemble. (**b**) Distribution of start and end configurations for trajectories in each ensemble. (**c**) Distribution of trajectory durations for trajectories in each ensemble

transition paths and trajectories generated from protocol simulations by projecting the state space onto the committor coordinate.

The trajectory ensemble generated by protocol simulations and the transition-path ensemble differ in the duration of trajectories and the configuration at the beginning and end. In the TPE, the endpoint configurations of each trajectory are the same by definition (the all-down and all-up configurations, respectively), but the duration of each trajectory differs. In contrast, each trajectory generated by the protocol is of equal duration, but the start and end configurations are not constrained, as discussed below.

The difference in endpoints between the processes can be approximately resolved by choosing control parameters and protocol endpoints that restrain the system to configurations corresponding to the reaction endpoints. For the Ising model, the reactant A is defined to be the all-down configuration and the product B to be the all-up configuration; these configurations have committor values $q(\boldsymbol{\phi} \in A) = 0$ and $q(\boldsymbol{\phi} \in B) = 1$, respectively. For the Ising model, the 4D fully optimized protocol starts with all fields at $\lambda_{i,\text{initial}} = -0.5k_\text{B}T$ and ends with $\lambda_{i,\text{final}} = 0.5k_\text{B}T$, whose equilibrium distributions are dominated by the all-down or all-up configuration, respectively (which leads to mean system magnetization close to -1 and 1). The system does not reach the equilibrium distribution by the end of the protocol, but for sufficiently long protocols will be "close" to this endpoint. The system's distance from the equilibrium distribution depends on the protocol duration, but since minimum-work protocols in the linear-response regime are independent of protocol duration, a single protocol duration ($\Delta t = \tau_\text{rel}$) in this regime is used as a representative of the process for all protocols in the linear-response regime. Figure 5.2b shows that the distribution of committor values for configurations at the start and end of the protocol performed in duration $\Delta t = \tau_\text{ref}$ are highly peaked in the $q = 0$ and $q = 1$ states respectively, indicating that on average the trajectories generated by a protocol of this duration transit between the reaction endpoints.

Resolving the range of transition-path durations requires a different approach. Because the system must wait for appropriate fluctuations from the environment to make progress during a transition from all down to all up, each step requires a stochastic amount of time and the overall transition paths have a range of durations. Additionally, the system can move "backward" during the trajectory, leading to several recrossings of a dividing surface between A and B. (Note that these are the same recrossing events that are ignored in transition-state theory and minimized in variational transition-state theory through optimization of the dividing surface). Combining information from multiple trajectories is therefore challenging, as at a given time t since the start of the transition path, one trajectory could be significantly closer to the product than another. This contrasts with control protocols where the external fields are the same at time t for all trajectories, and for long-duration protocols, the system state will be close to equilibrium and therefore relatively consistent across multiple trajectory samples.

If the reactive trajectories are described by a one-dimensional coordinate $x(t)$ (for example, from single-molecule force spectroscopy [4]), there are methods to combine temporal information from trajectories of multiple durations to obtain a mean transition path time for each value of the one-dimensional coordinate, $\langle t(x) \rangle$, using various ideas based on mean first-passage times [5–8]. However, an average one-dimensional trajectory is not of interest to the present analysis since general mechanistic information is desired in terms of physically intuitive collective variables (here given by each spin's magnetization).

Two approaches are therefore used for averaging information from the transition-path ensemble, one that aligns the time axis of each trajectory and one that aligns committor values. The time in each trajectory can be linearly rescaled by its total duration τ_{traj} to get a reaction coordinate $t/\tau_{\text{traj}} \in [0, 1]$ that represents the proportion of time elapsed during the trajectory, which is termed the *scaled transition-path time*. In practice, the scaled transition-path time t/τ_{traj} is divided into 20 discrete bins and the mean properties in each bin for each trajectory is determined, then averaged over all trajectories, ensuring that long trajectories are not disproportionately weighted in the ensemble average [7]. This is a relatively *ad hoc* approach to rescaling transition paths of multiple durations that would benefit from further improvement, but it nevertheless captures mechanistic information.

Another approach uses the committor, which in principle contains the information about the reaction mechanism in the transition-path ensemble since it is the "ideal" reaction coordinate. For a continuous system, visualizing the isocommittor surfaces throughout a collective-variable space would provide information about the order of events and the bottle-necks of the reaction. However, the Ising model has a discrete state space and there is no guarantee that a particular isocommittor surface is populated by any system configurations (for example, there is no configuration with $q(\boldsymbol{\phi}) = 0.5$). Instead, the range of committor values is divided into discrete bins and all configurations from each trajectory that fall in the same bin are grouped together. Such an approach allows calculation of the mean properties of each spin during the transition path. Averaging over the committor coordinate directly tends to under-emphasizes configurations at the start and end of the transition mechanism, and therefore a nonlinear transformation of the committor is used to make these configurations more visible. Inspired by the mathematical form of TPE entropy production in Chap. 3, the committor is transformed by the mapping

$$f(\boldsymbol{\phi}) = \ln \frac{q(\boldsymbol{\phi})}{1 - q(\boldsymbol{\phi})}, \tag{5.4}$$

and configurations that fall into 9 evenly spaced bins for this coordinate are grouped together. Note that an invertible transformation of the committor does not affect the information contained in the one-dimensional coordinate, and therefore $f(\boldsymbol{\phi})$ is still a "perfect" reaction coordinate.

5.3 Reaction Mechanism for Spin Inversion

Figure 5.3 shows the mean state of each spin set during the 4D fully optimized protocol and during the transition path as parameterized by the scaled committor or scaled transition-path time. The two methods of averaging the mean state in the transition-path ensemble look very similar, but using scaled transition-path time allows one to use more discrete bins and obtain smoother curves. The mean state of the system during the 4D fully optimized protocol and during the transition-path ensemble show similar characteristics, but differ in their finer details. In general, the green spins (which are initially energetically frustrated due to the adjacent boundary spins of opposite sign) flip relatively early in both processes, and the blue spins (which end in an energetically frustrated orientation) flip relatively late. The red and black spins flip throughout the middle of the protocol. However, the finer details of the processes differ; the MWP seems to flip both green spins around the same time (early in the protocol), while in the TPE at least one red spin flips relatively early, on average before the second green spin.

The differences in the detailed mechanisms may be due to the constraint on driving in the 4D control parameter space where, e.g., all four red fields are changed in the same way throughout the protocol. This constraint means it is not possible for the protocol to drive one spin to the up orientation independently of other spins in the same spin set. The resulting protocol in the 4D space is therefore symmetric in the identity of each of the spins. In contrast, trajectories in the transition-path ensemble seem to break this symmetry, but the average state of each spin as a function of the rescaled committor or scaled transition-path time does not show this symmetry-breaking.

Therefore, a method is developed to determine mechanistic information that distinguishes between spins in the same spin set from the trajectories generated by MWPs or that form the TPE. Since the order of spin flips during the mechanism is of interest, the order l_α of spin flips (with $\alpha = 1 \ldots 9$) is determined for each trajectory that gives the identity of the α-th spin to flip during a trajectory. Each spin may flip

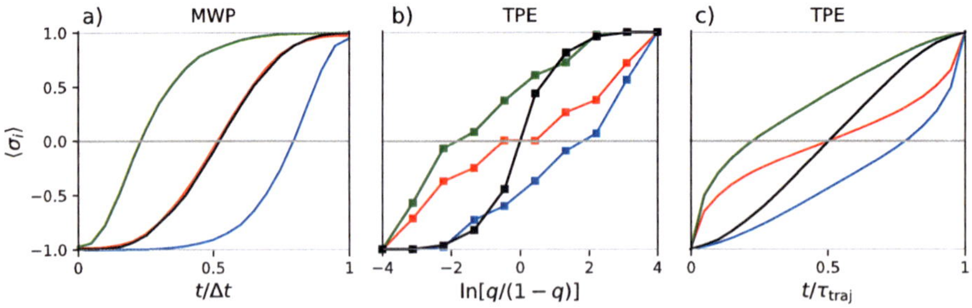

Fig. 5.3 Mean magnetization of each spin set (colors in Fig. 5.1) (**a**) at scaled time $t/\Delta t$ during the minimum-work protocol, (**b**) binned to the rescaled committor $\ln[q/(1-q)]$ during the transition-path ensemble, and (**c**) at scaled transition-path time t/τ_{traj} during the transition-path ensemble

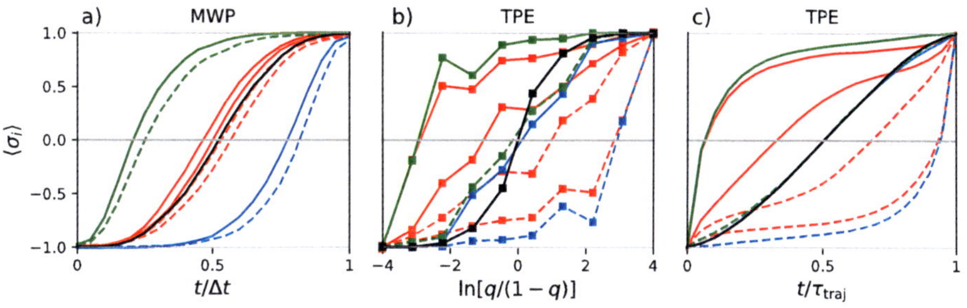

Fig. 5.4 Symmetry-broken mechanism for the spin-inversion transition. The mean state of each spin during the (**a**) MWP, (**b**) TPE as a function of the scaled committor, and (**c**) TPE as a function of scaled trajectory time t/τ_{traj}. The first spin(s) to flip in each spin set are indicated by solid curves and the last spin(s) are indicated by dashed curves

multiple times in a single trajectory, so to get an average order for the trajectory, the proportion of time spent in the up orientation for each spin is calculated, then all spins are ordered in descending proportion: a spin that spends more time during the trajectory in the up orientation than another is identified as flipping earlier in the reaction (has a lower value of α). There are four spin types identified by colors in Fig. 5.1. To break this symmetry, nine spin types are identified: the first, second, third, and fourth red spin; the first and second green spin; the first and second blue spin; and the black spin. These coarse variables still do not contain exact geometric information (e.g., they do not specify where the second red spin is located relative to the first red spin), but do break the symmetry within a single trajectory. The same analysis is performed on all trajectories in the ensemble to break the symmetry between spins in the same spin set. Once the order of each spin in the trajectory is determined so that spins in the same spin set are ordered, the mean magnetization of each spin type can be averaged over all trajectories.

Figure 5.4 shows the average trajectory during the MWP and TPE with broken symmetry for spins in the same spin set. In the MWP, each spin-flip event happens at relatively distinct times during the protocol: both green spins flip early, then all red and black spins flip, then finally both blue spins. Spins in the same spin set flip at approximately the same time, reflecting the symmetry of the driving protocol. In contrast, the TPE shows symmetry breaking for spins in the same spin set: a green and red spin flip first; followed by a second red spin; then the second green, first blue, and black spins; then the third red spin; and finally the fourth red and second blue spins. Such symmetry breaking cannot be achieved by a 4D protocol, therefore the mechanism observed in the 4D MWP cannot precisely match the mechanism observed in the TPE. However, the general feature of green spin(s) flipping early and blue spin(s) flipping late is preserved in both processes, reflecting the bias imposed by the fixed boundary spins that affects both the geometry of control-parameter space and the spontaneous transition mechanism.

While Fig. 5.4 separates spin flips in the same spin set, it does not show the overall order of spin flips during the trajectory. This is shown in Fig. 5.5 by plotting

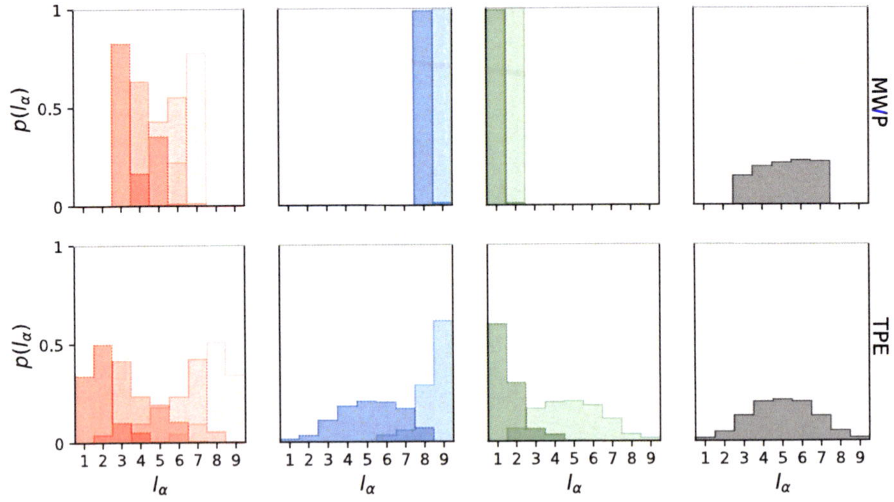

Fig. 5.5 Distribution of spin-flip order l_α for each spin type during the MWP (top) and TPE (bottom). Each histogram shows the distribution of the spin-flip order for a given spin, with spins in the same set separated based on their flipping order in the trajectory. The earlier spin(s) of each spin type have darker shading in the histogram, while later spin(s) are lighter

the distribution of the spin-flip order l_α of each spin type relative to all others. The MWP preserves the order across all trajectories, with both green (blue) spins flipping first (last), and red and black spins all flipping in between. The TPE spin-order distributions, however, show higher diversity. The first red and first green spins predominantly flip first, and the second blue and last red predominantly flip last, but the order of other spins is more variable. This indicates that there are many paths taken (with relatively high probability) by the system with spin flips occurring in different orders. It is possible that the TPE transition mechanism in this system cannot be characterized by a single ordering of spin flips, and instead some other physically intuitive collective variables that are able to capture the variety of transition paths would be better suited to describing the mechanism.

5.4 Energy Flows During the Minimum-Work Protocol and Transition-Path Ensemble

Control protocols produce changes in the system configuration by doing work. In the quasi-static limit, a protocol drives the system over an internal energy barrier by first providing some of this energy as work to increase the system's internal energy and drive it over the barrier, then extracting work from the system as it decreases in energy. Transition paths overcome an energy barrier by waiting for a sequence of fluctuations from the environment that provide the necessary energy as heat to the system. It seems intuitive that if the system must overcome some internal energy barrier while changing between collective-variable endpoints, then a "good" control protocol would provide this energy to the system as work rather than waiting for the appropriate (rare) thermal fluctuations from the environment, essentially flattening the total energy landscape along the path the system takes through configuration space. Here, the energy flows into the system during each trajectory are determined and split between motion in each spin. The idea of analyzing energy flows in the transition-path ensemble has been discussed previously [1, 2] where energy flows are shown to be helpful for determining collective variables that may be relevant to the reaction.

During a transition path, no work is done on the system and therefore changes in internal energy during the trajectory is due to heat flow. For a trajectory $\{\boldsymbol{\phi}(t_k)\}$ with $k = 1 \ldots \tau_{\text{traj}}$, the total heat flow to the system is

$$Q(\tau_{\text{traj}}) = \sum_{k=1}^{\tau_{\text{traj}}-1} [E_{\text{int}}(\boldsymbol{\phi}(t_{k+1})) - E_{\text{int}}(\boldsymbol{\phi}(t_k))] = E_{\text{int}}(B) - E_{\text{int}}(A) , \tag{5.5}$$

which is equal to the difference in internal energy of B relative to A. For the Ising system, $\Delta E_{AB} = E_{\text{int}}(B) - E_{\text{int}}(A) = 0$ due to the underlying symmetry of the metastable states. Since the single-spin-flip dynamics that are implemented are multipartite (a multidimensional extension of the bipartite dynamics condition in Sect. 3.3 where only one coordinate moves in a given time step while all others remain stationary), the total heat flow can be split between each spin to better understand the energetic costs associated with each spin's dynamics during the mechanism. The heat flow to spin σ_i during the trajectory is

$$Q^i(\tau_{\text{traj}}) = \sum_{k=1}^{\tau_{\text{traj}}-1} [E_{\text{int}}(\boldsymbol{\phi}(t_{k+1})) - E_{\text{int}}(\boldsymbol{\phi}(t_k))]\{1 - \delta[\sigma_i(\boldsymbol{\phi}(t_{k+1}) - \sigma_i(\boldsymbol{\phi}(t_k))]\} \tag{5.6a}$$

$$= \sum_{k=1}^{\tau_{\text{traj}}-1} \Delta E^i_{\text{int}}(k \to k+1) \tag{5.6b}$$

where the final factor in Eq. (5.6a) ensures that only transitions where σ_i changes contribute to the average. The mean heat flow to σ_i is binned to either the scaled transition-path time or rescaled committor:

$$Q^i(t/\tau_{\text{traj}}) = \sum_{k=1}^{\tau_{\text{traj}}-1} \Delta E^i_{\text{int}}(k \to k+1)\, \delta\left[\frac{t}{\tau_{\text{traj}}} - \frac{k+0.5}{\tau_{\text{traj}}}\right] \tag{5.7a}$$

$$Q^i(f) = \sum_{k=1}^{\tau_{\text{traj}}-1} \Delta E_{\text{int}}^i(k \to k+1)\, \delta\left[f - \frac{f(\boldsymbol{\phi}(t_{k+1})) + f(\boldsymbol{\phi}(t_k))}{2}\right], \tag{5.7b}$$

where the condition $\delta[t/\tau_{\text{traj}} - (k+0.5)/\tau_{\text{traj}}]$ selects all transitions that fall in the discrete time bin centered at t/τ_{traj}, and $\delta[f - \frac{1}{2}(f(\boldsymbol{\phi}(t_{k+1})) + f(\boldsymbol{\phi}(t_k)))]$ selects all transitions that fall in the discrete rescaled committor bin centered at $f = \ln[q/(1-q)]$. The heat flow to spin σ_i can then be averaged over all trajectories to obtain the ensemble-average heat flow $\langle \dot{Q}_{\text{R}}^i \rangle$ to that spin during the trajectory. If heat flow is positive into a particular σ_i, i.e. $\langle \dot{Q}_{\text{R}}^i \rangle > 0$, this suggests that motion in σ_i primarily "activates" the system towards the energy barrier, while $\langle \dot{Q}_{\text{R}}^i \rangle < 0$ implies that motion in σ_i primarily relaxes the system to the product [1].

During the control protocol, both work and heat change the total energy of the system. The work and heat steps during the trajectory are considered separately. Work is identified as the change in total energy $V(\boldsymbol{\phi}, \boldsymbol{\lambda})$ when control parameters change at fixed system position,

$$\mathcal{W}(\Delta t) = \sum_{k=1}^{\Delta t-1} [V(\boldsymbol{\phi}(t_k), \boldsymbol{\lambda}(t_{k+1})) - V(\boldsymbol{\phi}(t_k), \boldsymbol{\lambda}(t_k))] \tag{5.8a}$$

$$= -\sum_{k=1}^{\Delta t-1} \sum_i [\lambda_i(t_{k+1}) - \lambda_i(t_k)] X_i(t_k) \tag{5.8b}$$

$$= \sum_i \mathcal{W}^i(\Delta t), \tag{5.8c}$$

which can be partitioned among the control parameters λ_i. Heat is the change in total energy when the system state changes for fixed control parameters

$$Q(\Delta t) = \sum_{k=1}^{\Delta t-1} \left[V\big(\boldsymbol{\phi}(t_{k+1}), \boldsymbol{\lambda}(t_{k+1})\big) - V\big(\boldsymbol{\phi}(t_k), \boldsymbol{\lambda}(t_{k+1})\big)\right] \tag{5.9a}$$

$$= \sum_{k=1}^{\Delta t-1} \Big(\big[E_{\text{int}}\big(\boldsymbol{\phi}(t_{k+1})\big) - E_{\text{int}}\big(\boldsymbol{\phi}(t_k)\big)\big] - \sum_i \lambda_i(t_{k+1})\big[X_i(t_{k+1}) - X_i(t_k)\big]\Big) \tag{5.9b}$$

$$= \sum_{k=1}^{\Delta t-1} \sum_i \Big(\big[E_{\text{int}}\big(\boldsymbol{\phi}(t_{k+1})\big) - E_{\text{int}}\big(\boldsymbol{\phi}(t_k)\big)\big]\{1 - \delta[\sigma_i(t_{k+1}) - \sigma_i(t_k)]\}$$
$$- \lambda_i(t_{k+1})\big[X_i(t_{k+1}) - X_i(t_k)\big]\Big) \tag{5.9c}$$

$$= \sum_{k=1}^{\Delta t-1} \sum_{i} \left[\Delta E^{i}_{\text{int}}(k \to k+1) + \Delta E^{i}_{\text{ext}}(k \to k+1)\right], \tag{5.9d}$$

where in Eq. (5.9d) heat flow due to changes in σ_i is identified as making distinct changes to the internal and external energy. Together, the work and heat change the total energy,

$$\Delta V(\Delta t) = \mathcal{W}(\Delta t) + Q(\Delta t) \tag{5.10a}$$

$$= \sum_{i} \left[\mathcal{W}^{i}(\Delta t) + Q^{i}(\Delta t)\right] \tag{5.10b}$$

which is split into contributions from motion in each control parameter and associated spin magnetizations. Motion in control parameters is deterministic, while motion in spins is stochastic.

Control protocols and transition paths share the internal energy landscape E_{int}. Therefore changes in internal energy during both processes will be analyzed to better understand how the system overcomes the energy barrier. After identifying the order of spin flips in a trajectory, the energy flow associated with each spin in the trajectory is calculated. This allows identification of differences in energy flows to spins in the same spin set that may have different roles in the reaction mechanism.

Figure 5.6 shows the mean heat flow to the system during the minimum-work protocol and during a transition path. When the system moves during the MWP, some of the heat flow changes internal energy and some of it changes external energy. The change in internal energy (Fig. 5.6b) for the green spins is positive during the first half of the protocol, for the blue spins is negative during the second half of the protocol, and for red and black is zero throughout the protocol. The changes in internal energy for the blue and green spins are compensated by changes in the external energy (Fig. 5.6a): the external energy decreases on average when green spins flip (increasing the internal energy) and increases on average when blue spins flip (decreasing internal energy). This demonstrates how external energy provides a force to the system which allows it to access high-internal-energy configurations during the protocol. When control parameters do work they change only the external energy; therefore this biasing force that allows the system to access high-internal-energy configurations is provided by control parameters. Additionally, spins from the same spin set have nearly identical energy flows during the protocol, again reflecting the symmetry of the control parameters.

Figure 5.6c and d shows the heat flow (change in internal energy) during the transition-path ensemble for both scaled transition-path time and re-scaled committor. The symmetry-breaking observed in the reaction mechanism is also reflected in the changes in internal energy during the transition path. The first green and first red spins take in energy on average early in the transition path, bringing the system out of the all-down configuration into higher-internal-energy configurations. Similarly, at the end of the transition path, the second blue and

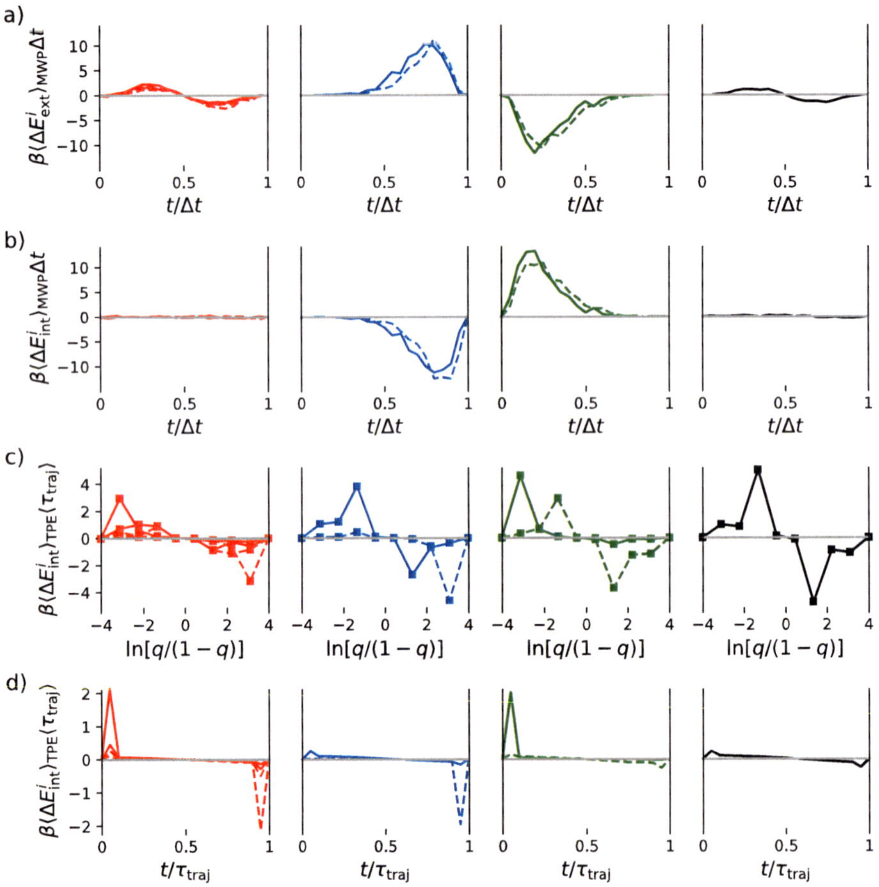

Fig. 5.6 Heat flow to each spin during the MWP and TPE. Top two rows: Mean change in (**a**) external energy $\beta\langle\Delta E^i_{\mathrm{ext}}\rangle$ and (**b**) internal energy $\beta\langle\Delta E^i_{\mathrm{int}}\rangle$ due to magnetization change in σ_i during the MWP, scaled by protocol duration Δt. Bottom two rows: Mean change in internal energy $\beta\langle\Delta E^i_{\mathrm{int}}\rangle$ due to magnetization change in σ_i during the TPE (**c**) projected onto the scaled committor and (**d**) as function of scaled trajectory time, scaled by the mean transition-path duration $\langle\tau_{\mathrm{traj}}\rangle$. The first spin(s) to flip in each spin set are indicated by solid curves and the last spin(s) are indicated by dashed curves

last red spin release energy as the system reaches the low-internal-energy all-up configuration. Throughout the middle of the protocol, the other spins on average take in or release heat. In general, if a spin flips in the first half of the trajectory, the heat flow is positive; if it flips in the second half of the trajectory, the heat flow is negative.

The heat flow during the transition-path ensemble also demonstrates differences in how the scaled transition-path time and rescaled committor provide information about the mechanism. While the mean state in these coordinates looks similar with either representation of transition paths (Fig. 5.4), it is much harder to visualize the heat flows using the scaled transition-path time than the rescaled committor. The heat flow into the system during the first step out of the stable all-down configuration and the heat flow out of the system during the final step into the all-up configuration inevitably occur in the first and last bins of the scaled transition-path time leading to highly peaked heat flow at these times. Throughout the middle of the transition path, heat flow is small for two reasons: first, some of the internal energy changes for transitions throughout the protocol are zero and thus do not contribute magnitude to the heat flow; second, linearly rescaling the transition-path time can result in spin flips occurring at a wide range of scaled transition-path times that are not necessarily aligned between different trajectories, which spreads out the heat flow associated with this transition over multiple scaled trajectory-time bins. For this reason, the rescaled committor is a more desirable reaction coordinate, since it provides a measure of the system's progress between reaction endpoints independent of the time that the system reaches a state in any given trajectory. However, qualitative aspects of the heat flows for the two coordinates are the same.

5.5 Summarizing the Mechanisms

Finally, the reaction mechanisms identified for the minimum-work protocol and the transition-path ensemble are summarized.

The minimum-work protocol has a clear and intuitive reaction mechanism, shown in Fig. 5.7. In the first two steps, both green spins are flipped. Each of the green spins breaks three energetically favourable interactions and forms one when it flips. The change in internal energy from each interaction has an absolute value of $2k_{\mathrm{B}}T$ for a total change of $4k_{\mathrm{B}}T$. This positive heat flow to green spins is also shown in Fig. 5.6. Once these spins have flipped, the four red spins and black spin can flip with no change in internal energy, since they each have two neighbouring spins in the up orientation and two in the down orientation. They each experience a driving force from the changing magnetic fields which is equivalent across all of these spins, leading to broad distributions over step number for all of these spins (Fig. 5.5). In the final two steps the blue spins flip, each of which releases $4k_{\mathrm{B}}T$ of internal energy as the system relaxes into the all-up configuration (Fig. 5.6).

Although the details are harder to pin down, Fig. 5.8 shows a mechanism for spin-inversion during a spontaneous transition path. The first two steps flip a red and green spin (which can happen in either order) and take in $4k_{\mathrm{B}}T$ of heat in the first step. The next five steps can happen in many ways, and different spins will exchange heat with the environment depending on how the transition occurs. In general, one of these five steps will take in $4k_{\mathrm{B}}T$ of heat and one will release $4k_{\mathrm{B}}T$ of heat, and other steps can be accomplished with no change to the internal energy. The final two steps flip the last red and blue spins, decreasing system energy by $4k_{\mathrm{B}}T$ as it reaches the all-down configuration.

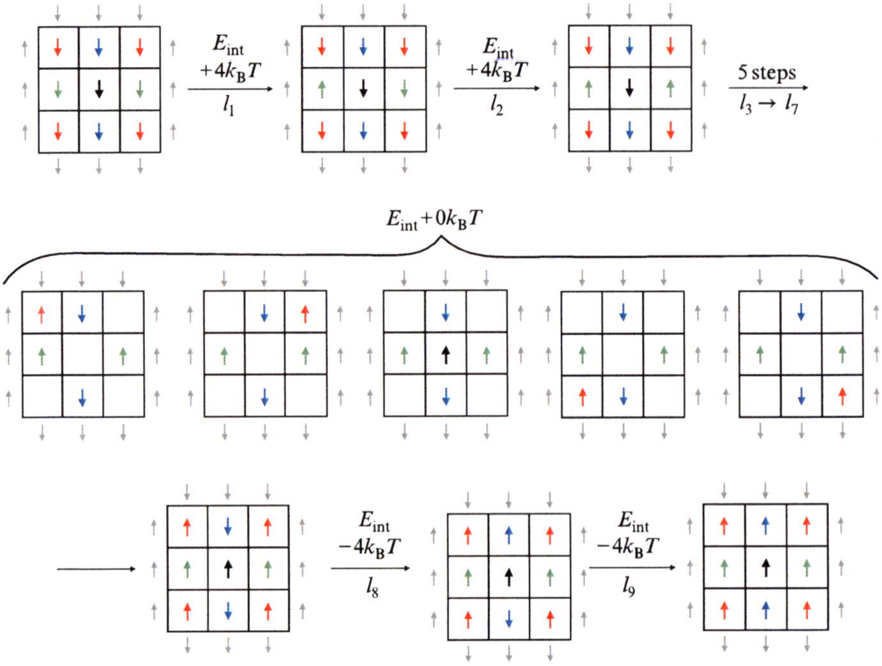

Fig. 5.7 Transition mechanism during the designed protocol (MWP). The change in internal energy and step number l_α are indicated above and below the arrows. The five middle steps can happen in any order, but each has no change in internal energy

5.6 Discussion

The transition mechanism for spin inversion in a 3×3 Ising model during the fully optimized 4D protocol and during a spontaneous transition path has been investigated. During the MWP, work is done on the system, which provides an energetic bias that drives the system up and over the energy barrier. In the transition-path ensemble, the system must wait for appropriate fluctuations of heat from the environment to overcome the same barrier.

The mechanisms contained in trajectories generated by each process were examined in detail, where a mechanism in this context is a sequence of spin-flip events. The MWP shows a clear transition mechanism, with both green spins taking in energy as they flip first, red and black spins flipping throughout the middle of the protocol with minimal energetic cost, and finally the blue spins flipping and releasing energy. This ordering preserves the symmetry of different spin types and is well-preserved across all trajectories.

The transition-path ensemble shows a wider diversity of transition paths that break the underlying symmetry of spin types. The first two spin flips are usually one green and one red spin, taking in $4k_BT$ of energy in the first flip and $0k_BT$ in the

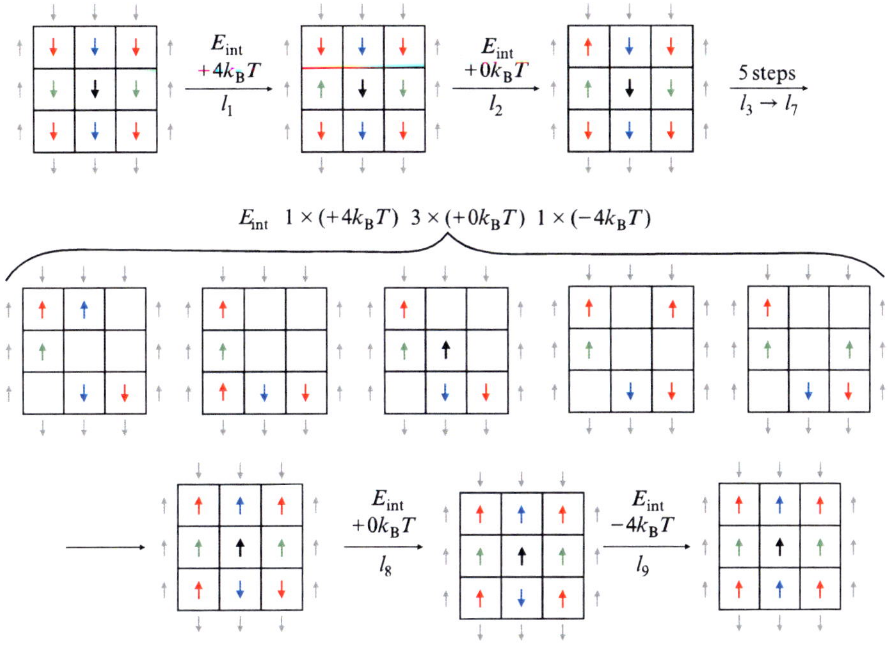

Fig. 5.8 Transition mechanism during the transition-path ensemble (TPE). The change in internal energy and step number l_α are indicated above and below the arrows. The five middle steps can happen in many orders, and one step will increase the internal energy by $4k_BT$ while another will decrease it by $4k_BT$

second. The next five steps have many orders of spin flips where at least one takes in $4k_BT$ of heat and another releases $4k_BT$. The final two steps involve flipping a red and blue spin, releasing $4k_BT$ of heat in the final step.

The internal energy barrier of $8k_BT$ is consistent between the two processes, but the overall manner in which the system overcomes the barrier differs. A significant factor affecting the comparison is the constrained symmetry of the 4D protocols, which prevents the controller from pushing on one spin differently than its symmetric counterpart(s). The transition-path ensemble shows symmetry-breaking between spins in the same set, both in the order in which spin flips occur during the mechanism and the energetic cost. Relieving this constraint requires optimizing a higher-dimensional protocol, which may show the symmetry-breaking required to generate a mechanism consistent with the transition path. The minimum-work protocol in 9D space is too computationally expensive to compute using the same methods used to generate the 4D fully optimized protocol. Minimum-work protocols using the linear-response approximation in higher-dimensional spaces have been computed (e.g., a 100D protocol in Ref. [9]), but under the assumption that relaxation time is constant throughout control-parameter space so that in the optimization the friction matrix can be approximated with the force covariance matrix, significantly reducing computational cost.

More generally, what can be said about connections between minimum-work protocols and spontaneous transition mechanisms? The choice of control parameters and their ability to affect change in collective variables that are relevant to describing the reaction mechanism is an important factor in the comparison. The 4D protocol studied here does not have the ability to reproduce the dynamics in the transition-path ensemble because some of the relevant information is coarse-grained. The Ising model has $2^9 = 512$ configurations, but only 168 are unique with respect to the symmetry operations. There are also 168 unique committor values, indicating that the specific geometry of each state is relevant to parameterizing the committor. The 4D conjugate-force vector, on the other hand, only distinguishes $5 \times 3 \times 3 \times 2 = 90$ unique configurations. Thus there is loss of information about the transition mechanism in the 4D collective-variable space. Choosing control parameters that drive all collective variables that are relevant to the reaction would allow a closer comparison with the transition-path ensemble.

The underlying transition mechanisms and heat flows differ between the processes. The MWP seems to drive the system through a specific set of configurations, providing a clear and interpretable transition mechanism and analysis of energy flow to the system. In contrast, for this system the transition-path ensemble has a variety of transition mechanisms which makes it difficult to determine the details of any one dominant mechanism. If the transition-path ensemble were such that reactive trajectories tend to follow the same path (i.e., lie in a "transition tube" [10]) and control parameters are chosen to push on the relevant collective variables to describe the transition, a stronger correspondence between the processes may be observed.

References

1. Li, W., Ma, A.: Reaction mechanism and reaction coordinates from the viewpoint of energy flow. J. Chem. Phys. **144**, 114103 (2016)
2. Wu, S., Li, H., Ma, A.: A rigorous method for identifying a one-dimensional reaction coordinate in complex molecules. J. Chem. Theory Comput. (2022)
3. Vanden-Eijnden, E.: Transition path theory. In: Bowman, G.R., Pande, V.S., Noe, F. (eds.) An Introduction to Markov State Models and Their Application to Long Timescale Molecular Simulation, chap. 7, pp. 91–100. Springer, Berlin (2014)
4. Hoffer, N.Q., Neupane, K., Pyo, A.G.T., Woodside, M.T.: Measuring the average shape of transition paths during the folding of a single biological molecule. Proc. Natl. Acad. Sci. **116**(17), 8125–8130 (2019)
5. Kim, W.K., Netz, R.R.: The mean shape of transition and first-passage paths. J. Chem. Phys. **143**(22) (2015)
6. Makarov, D.E.: Shapes of dominant transition paths from single-molecule force spectroscopy. J. Chem. Phys. **143**(19), 2–6 (2015)
7. Cossio, P., Hummer, G., Szabo, A.: Transition paths in single-molecule force spectroscopy. J. Chem. Phys. **148**, 123309 (2018)
8. Park, S., Sener, M.K., Lu, D., Schulten, K.: Reaction paths based on mean first-passage times. J. Chem. Phys. **119**(10), 1313 (2003)

9. Rotskoff, G.M., Crooks, G.E., Vanden-Eijnden, E.: A geometric approach to optimal nonequilibrium control: minimizing dissipation in nanomagnetic spin systems. Phys. Rev. E **95**, 012148 (2017)
10. Vanden-Eijnden, E., Venturoli, M.: Revisiting the finite temperature string method for the calculation of reaction tubes and free energies. J. Chem. Phys. **130**(19), 194103 (2009)

Chapter 6
Conclusions

Abstract This chapter summarizes the work in this thesis and provides suggestions for future investigation of the connection between efficient protocols and spontaneous transition mechanism, outlining general directions of investigation and particular model systems.

The study of reaction mechanisms is fundamental to the understanding of chemical and biological processes. Determination of physically intuitive reaction coordinates and the overall characterization of the reaction pathway from trajectories with microscopic detail is a central challenge in chemical physics today. As the ability to study larger systems with finer spatial and longer temporal resolution grows, there is greater need for theoretical frameworks that provide a basis for extraction and interpretation of reaction mechanisms in arbitrary systems. The development of transition-path theory [1, 2] over the last couple of decades provides this versatile theoretical description of a reaction through definition of an "ideal" reaction coordinate, the committor.

In Chap. 3, an information-theoretic interpretation of transition-path theory and the committor's ability to identify relevant coordinates is described. Information theory provides a powerful and versatile framework for interpreting many forms of data and the information it contains. In the context of reaction mechanisms, the committor is the ideal reaction coordinate because it preserves the mutual information (a statistical measure of dependency between random variables) between the system's microstate and the outcome (A or B) of the current trajectory segment under elimination of all other degrees of freedom. This satisfies the intuition about what a reaction coordinate is (a one-dimensional complete representation of a reactive event [3–5]) while providing a concrete understanding of the information preserved under projection of dynamics onto the reaction coordinate, in this case, information about the trajectory variables that define the transition-path ensemble and dictate its thermodynamics. Furthermore, incorporating information theory into the definition of the reaction coordinate provides a thermodynamic interpretation of the coarse-graining of system microstates that is done when low-dimensional collective variables are found to describe the reaction. In stochastic thermodynamics,

M. Louwerse, *Efficient Control and Spontaneous Transitions*, Springer Theses,
https://doi.org/10.1007/978-3-031-40534-1_6

a particular coarse-grained representation of system state is said to be thermodynamically consistent when the (physical) entropy production of the process is the same when calculated in the full state space and in the coarse-grained space [6, 7]. Low-dimensional collective variables that describe reaction dynamics should be thermodynamically consistent with the high-dimensional dynamics, but a definition of what this thermodynamic consistency entails has been lacking. The transition-path ensemble entropy production introduced in Sect. 3.2—that characterizes the time-asymmetry of the reaction dynamics—provides a definition of thermodynamic consistency for reaction dynamics. Additionally, a thermodynamic definition is provided for the as-yet vague concept of relevance of a collective variable to the reaction, which provides a criterion for optimization. This contribution provides a theoretical basis for the development of computational methods for extracting and interpreting information about the reaction mechanism using tools from information theory and stochastic thermodynamics.

In Chap. 4, minimum-work protocols for spin-inversion in a 2D Ising model are calculated and characterized, focusing on how multidimensional protocols interact with specific system coordinates and the underlying energy landscape to speed system relaxation and reduce excess work for the protocol. Designed multidimensional protocols take advantage of their flexibility to avoid control-parameter regions of high friction (Fig. 4.2a), decorrelate spins (Fig. 4.7) and flatten the energy landscape, thereby boosting the system population in configurations with high internal energy (Fig. 4.8), and heterogeneously input and extract work in concert with system relaxation timescales (Fig. 4.9). This drives the system along a fast-relaxing path connecting the configuration endpoints, keeping the system closer to equilibrium and reducing resistance and hence work relative to naive protocols. This work highlights the advantages of using several collective variables to drive the system since there is greater flexibility in choosing regions of configuration space to drive the system through. The designed protocols highlight some relevant features of the system's energy landscape, in particular the asymmetric boundary conditions on the system and (approximately) the size of the energy barrier the system must overcome during the reaction.

Finally, in Chap. 5 the transition-path ensemble for spin-inversion in the same 2D Ising model is studied, with methods developed to extract the reaction mechanism from trajectories and compare it to the mechanism during the 4D fully optimized protocol calculated in Chap. 4. The mechanisms differ in their finer details but the general features are preserved given the constrained symmetry of the 4D protocol. This highlights the importance of identifying relevant collective variables for the reaction: since the spontaneous transition mechanism depends on energetic and structural relationships between collective variables, control parameters that cannot affect change in the relevant collective variables may not be able to drive the system through regions of configuration space that are typically visited during a spontaneous transition. During a spontaneous transition in the 2D Ising model, spins of the same symmetry type (e.g. the two green spins) have different heat flow, and therefore the 4D control parameters that treat all spins of the same type equally cannot do work on individual spins in a way that matches their energetic requirements in the transition mechanism. However, given the constraint on the control parameters, the 4D fully optimized protocol drives the system along a path with a minimal internal energy barrier. It would be interesting to see how minimum-work protocols for this system designed in a 9D control parameter space compare to the 4D protocol and the spontaneous transition mechanism, but clever methods for optimizing the protocol in the high-dimensional space need to be developed. Nevertheless, the results suggest that minimum-work protocols are able to uncover aspects of the spontaneous transition mechanism, potentially providing another optimization criterion for finding relevant collective variables and spontaneous transition mechanisms from nonequilibrium driving protocols.

6.1 Future Directions

There are several avenues of investigation opened up by work in this thesis.

The information thermodynamics of transition paths derived in Chap. 3 opens several opportunities for future work. First, it would be helpful to better understand the meaning of the transition-path ensemble entropy production rate and how it depends on underlying details of the system, such as the magnitude of the energy barrier separating metastable states, position-dependent diffusion throughout configuration space, the presence of multiple distinct transition pathways, and the precise choice of reaction endpoints. To this end, it would be helpful to derive these results in a continuous state space to complement the Master-equation description presented in Chap. 3 and expand the applicability of the theory. Further, developing transition-path thermodynamics under coarse-graining of the state space [7] would be beneficial to more precisely understand the reaction information available from observation of a one-dimensional coordinate. Additionally, the Fisher information metric in configuration space defined in Sect. 3.2.3 is intriguing and may be worth further investigation using ideas from information geometry and its relationship to stochastic thermodynamics [8, 9]. With a better understanding of the intuition that can be gained about a reaction from its entropy production rate, single-molecule experiments examining transition-path dynamics could be performed and analyzed to yield insight into the reaction [10, 11].

The connection established between transition-path theory and stochastic thermodynamics suggests that relevant coordinates are identified by their time-asymmetry in the transition-path ensemble. A promising avenue for future work is the development of algorithms to find maximally informative coordinates from a sample of transition paths using established methodology for inferring entropy production in nonequilibrium systems [12–15]. Additionally, incorporating stochastic-thermodynamic ideas into the interpretation of reaction mechanisms could be insightful, by using the same concepts of work, heat, and information flows used to analyze the operation of molecular machines operating in a nonequilibrium steady state [16, 17]. In particular, analyzing energy flows into system coordinates was useful in interpreting the spin-inversion mechanism in

Chap. 5, and similar ideas have also been used to study conformational changes in alanine dipeptide [18, 19]. This analysis of energy flows goes beyond a one-dimensional reaction free-energy profile (Fig. 1.1) and provides more detailed understanding of the energy landscape and how it depends on collective variables of interest.

Another direction is to extend the information thermodynamics for a transition between A and B to describe a network of transition paths between N distinct subsets of configurations I. This has strong ties to Markov-state modeling [20] and milestoning [21], which use transition dynamics between nearby subsets of (metastable) configurations to describe dynamics on a global scale between distant configurations of interest. In this case, the trajectory outcome and origin would be multinomial random variables, with conditional probabilities given by some multinomial generalization of the committor. Each node would have its own $I \rightarrow I$ stationary subensemble with no entropy production, as well as a set of reactive subensembles going to each other node, each with its own entropy production. This gives a total of $\binom{N}{2}$ reactive subensembles for the entire system, in principle requiring $\binom{N}{2}$ reaction coordinates that generalize the traditional binomial committor. Nevertheless, the information generated in this multinomial scenario about the multivariate trajectory outcome would be equal to the average time-irreversibility throughout all these reactive subensembles. Relevant coordinates that maximize entropy production in a set of reactive subensembles of interest could be found, gaining insight into the overall reaction network and collective variables describing each transition.

Transition-path thermodynamics were derived for a system in equilibrium, but in principle this theory could be extended to nonequilibrium steady states, where the system is connected to one or more reservoirs that maintain it out of equilibrium, breaking the detailed balance condition which is used to derive several relationships in Chap. 3. In a nonequilibrium context, the mechanism for a system transition from a reactant to a product can also be considered, and the time-asymmetry of dynamics within a transition-path ensemble and the information generated by particular degrees of freedom could be derived separately from the thermodynamic entropy production of the underlying dynamics. This may have connections with recent work by Ref. [22], which showed that milestoning (a transition-path sampling technique [21] that essentially uses a multinomial version of the trajectory outcome and origin variables [23]) is a thermodynamically consistent method of coarse-graining nonequilibrium dynamics without losing information about entropy production. There are strong similarities to the ideas discussed in Chap. 3, and a more concrete connection between these ideas is another avenue for cross-pollination of ideas from the fields of stochastic thermodynamics and transition-path theory.

Regarding minimum-work protocols, there is a need for better computational methods to design minimum-work protocols, especially in multidimensional control-parameter spaces. There has been recent progress in machine-learning techniques aimed at finding minimum-work protocols [24] which do not use the generalized friction and can therefore provide minimum-work protocols for any protocol duration. However, the generalized friction and its role in defining thermodynamic geometry in control-parameter space is physically appealing and suggests deeper connections with the geometry of the configuration space it has access to, warranting further investigation into its properties. Therefore better methods to calculate the generalized friction and its geodesics in a multidimensional control-parameter space would be helpful in expanding the applications of this approach to designing efficient protocols.

Finally, methods to optimize control parameters for efficiently driving the system would be a useful avenue for further investigation. The work in this thesis suggests that small coordinate sets identified to be relevant to a reaction could be driven with minimal energetic cost by appropriately choosing control parameters to drive them. This could plug into the immense progress in using machine learning to learn small sets of collective variables to drive a system across energy barriers [25–32]. The correspondence between "good" collective variables and "good" control parameters (and particular forms of the energetic coupling between control parameters and collective variables) that do minimal work could be studied in a more diverse set of model systems to further elucidate connections between minimum-work control and spontaneous transitions.

6.2 Final Remarks

Reaction mechanisms are a fundamental concept in chemistry and biology, and yet their theoretical description is challenging. Concepts such as the transition state and a one-dimensional reaction coordinate provide a starting point for development of reaction theories, but lack *a priori* definition without first having some intuition about the reaction of interest. The development of transition-path theory [1, 2] has alleviated some of the ambiguity associated with these concepts through definition of the committor, which provides the one-dimensional reaction coordinate and identifies the transition-state ensemble as the $q = 0.5$ surface. Yet challenges remain in interpreting the committor and gaining physical intuition about the mechanism in terms of small sets of system coordinates.

In this thesis, it is shown that stochastic thermodynamics provides some of the remaining pieces of the puzzle, with concepts of entropy production, information generation, and energy flows allowing identification of relevant dynamics and interpretation of the reaction mechanism. Given the rapid theoretical developments in the field in the last few decades, it should not be surprising that this powerful theoretical framework, which combines kinetics and thermodynamics by defining thermodynamic quantities for individual trajectories, is well-suited for describing the transient dynamics and energetics of a system as it undergoes a reaction. Further, the collective motions of the reaction mechanism intricately depend on interactions between collective variables. Therefore the response to dynamic manipulation of these collective variables during a protocol reflects their interactions. Protocols that minimize work take advantage of the interactions between collective variables that the control parameters affect, providing work to collective variables in accordance with the energetic requirements of their motion and providing fast-relaxing paths through configuration space. While these concepts were studied in a particular model system, the concepts and resulting intuition are general enough to be applied to other systems. The work in this thesis provides an initial road map for further investigations of the connections between efficient control and spontaneous transitions.

References

1. Weinan, E., Vanden-Eijnden, E.: Towards a theory of transition paths. J. Stat. Phys. **123**(3), 503–523 (2006)
2. Metzner, P., Schutte, C., Vanden-Eijnden, E.: Transition path theory for Markov jump processes. Multiscale Model. Simul. **7**(3), 1192–1219 (2009)
3. Peters, B.: Reaction coordinates and mechanistic hypothesis tests. Annu. Rev. Phys. Chem. **67**, 669–690 (2016)
4. Bolhuis, P.G., Dellago, C.: Practical and conceptual path sampling issues. Eur. Phys. J. Spec. Top. **224**, 2409–2427 (2015)
5. Li, W., Ma, A.: Recent developments in methods for identifying reaction coordinates. Mol. Simul. **40**(10–11), 784–793 (2014)
6. Seifert, U.: Stochastic thermodynamics, fluctuation theorems and molecular machines. Rep. Prog. Phys. **75**, 126001 (2012)
7. Esposito, M.: Stochastic thermodynamics under coarse-graining. Phys. Rev. E **85**, 041125 (2012)
8. Amari, S.-I.: Information Geometry and Its Applications, 1st edn. Springer, Tokyo (2016)
9. Ito, S., Oizumi, M., Amari, S.-I.: Unified framework for the entropy production and the stochastic interaction based on information geometry. Phys. Rev. Res. **2**, 33048 (2020)
10. Neupane, K., Hoffer, N.Q., Woodside, M.T.: Measuring the local velocity along transition paths during the folding of single biological molecules. Phys. Rev. Lett. **121**, 018102 (2018)
11. Pyo, A.G.T., Hoffer, N.Q., Neupane, K., Woodside, M.T.: Transition-path properties for folding reactions in the limit of small barriers. J. Chem. Phys. **149**, 115101 (2018)
12. Seifert, U.: From stochastic thermodynamics to thermodynamic inference. Annu. Rev. Condens. Matter Phys. **10**, 171–192 (2019)
13. Li, J., Horowitz, J.M., Gingrich, T.R., Fakhri, N.: Quantifying dissipation using fluctuating currents. Nat. Commun. **10**, 1666 (2019)
14. Skinner, D.J., Dunkel, J.: Improved bounds on entropy production in living systems. PNAS **118**(18), e2024300118 (2021)
15. Gnesotto, F.S., Gradziuk, G., Ronceray, P., Broedersz, C.P.: Learning the non-equilibrium dynamics of Brownian movies. Nat. Commun. **11**, 5378 (2020)
16. Lathouwers, E., Lucero, J.N., Sivak, D.A.: Nonequilibrium energy transduction in stochastic strongly coupled rotary motors. J. Phys. Chem. Lett. **11**(13), 5273–5278 (2020)
17. Leighton, M.P., Sivak, D.A.: Performance scaling and trade-offs for collective motor-driven transport. New J. Phys. **24**(1), 013009 (2022)
18. Li, W., Ma, A.: Reaction mechanism and reaction coordinates from the viewpoint of energy flow. J. Chem. Phys. **144**, 114103 (2016)
19. Wu, S., Li, H., Ma, A.: A rigorous method for identifying a one-dimensional reaction coordinate in complex molecules. J. Chem. Theory Comput. **18**(5), 2836–2844 (2022)
20. Chodera, J.D., Noé, F.: Markov state models of biomolecular conformational dynamics. Curr. Opin. Struct. Biol. **25**, 135–144 (2014)

21. Faradjian, A.K., Elber, R.: Computing time scales from reaction coordinates by milestoning. J. Chem. Phys. **120**, 10880–10889 (2004)
22. Hartich, D., Godec, A.: Emergent memory and kinetic hysteresis in strongly driven networks. Phys. Rev. X **11**, 041047 (2021)
23. Berezhkovskii, A.M., Szabo, A.: Committors, first-passage times, fluxes, Markov states, milestones, and all that. J. Chem. Phys. **150**, 54106 (2019)
24. Engel, M.C., Smith, J.A., Brenner, M.P.: Optimal control of nonequilibrium systems through automatic differentiation. arXiv:2201.00098
25. Wang, Y., Tiwary, P.: State predictive information bottleneck. J. Chem. Phys. **154**, 134111 (2021)
26. Ma, A., Dinner, A.R.: Automatic method for identifying reaction coordinates in complex systems. J. Phys. Chem. B **109**, 6769–6779 (2005)
27. Noé, F., Nüske, F.: A variational approach to modeling slow processes in stochastic dynamical systems. Multiscale Model. Simul. **11**(2), 635–655 (2013)
28. Pérez-Hernández, G., Paul, F., Giorgino, T., De Fabritiis, G., Noé, F.: Identification of slow molecular order parameters for Markov model construction. J. Chem. Phys. **139**(1), 015102 (2013)
29. Mardt, A., Pasquali, L., Wu, H., Noé, F.: VAMPnets for deep learning of molecular kinetics. Nat. Commun. **9**(1), 1–11 (2018)
30. Tiwary, P., Berne, B.J.: Spectral gap optimization of order parameters for sampling complex molecular systems. Proc. Natl. Acad. Sci. U. S. A. **113**(11), 2839–2844 (2016)
31. Wang, Y., Marcelo, J., Ribeiro, L., Tiwary, P.: Past-future information bottleneck for sampling molecular reaction coordinate simultaneously with thermodynamics and kinetics. Nat. Commun. **10**, 3573 (2019)
32. Hernández, C.X., Wayment-Steele, H.K., Sultan, M.M., Husic, B.E., Pande, V.S.: Variational encoding of complex dynamics. Phys. Rev. E **97**(6), 1–11 (2018)

Appendix A
Code

Select code supporting the work in this thesis is publicly available in a Github repository [1]. Further code and data are available upon reasonable request.

M. Louwerse, *Efficient Control and Spontaneous Transitions*, Springer Theses,
https://doi.org/10.1007/978-3-031-40534-1

Appendix B
Computational Details for Ising Model

B.1 Calculation of Generalized Friction Matrix

The friction matrix for control parameter $\boldsymbol{\lambda}$ was estimated from time-correlation functions of pairs of conjugate forces (2.62). Single-spin-flip Glauber dynamics [2] are used to simulate four equilibrium trajectories, each of length $10^7\,\mathrm{d}t$, where $\mathrm{d}t$ is a constant simulation time interval after which a random spin flip is attempted. Time-correlation functions were calculated between pairs of conjugate-force trajectories using fast Fourier transforms [3] and integrated to lag time $2.5 \times 10^4\,\mathrm{d}t$. The four independent estimates of friction at each control-parameter value were then averaged.

The 4×4 friction matrix is calculated at grid points with discrete spacing $\mathrm{d}\lambda_i = 0.2$ within the control-parameter space bounded by: $\lambda_{\mathrm{red}} \in [-2.1, 2.1]$, $\lambda_{\mathrm{blue}} \in [-3.0, 1.0]$, $\lambda_{\mathrm{green}} \in [-1.0, 3.0]$, $\lambda_{\mathrm{black}} \in [-2.1, 2.1]$. Since this control-parameter space has spatial symmetry, statistical power is doubled by averaging friction values for symmetrically equivalent control-parameter values: $(\lambda_{\mathrm{red}}, \lambda_{\mathrm{blue}}, \lambda_{\mathrm{green}}, \lambda_{\mathrm{black}}) \rightarrow (-\lambda_{\mathrm{red}}, -\lambda_{\mathrm{green}}, -\lambda_{\mathrm{blue}}, -\lambda_{\mathrm{black}})$.

B.2 Design of Control Protocols

Here, the methods used to design minimum-work protocols are described in more detail.

The time-optimized protocols are effectively one-dimensional with a single field applied equally to all spins; these protocols have optimal velocity inversely proportional to the square root of the 1D friction coefficient $\zeta^{\mathrm{1D}}(\boldsymbol{\lambda}) \equiv \sum_{i,j} \zeta_{ij}(\boldsymbol{\lambda})$.

The fully optimized protocols are calculated using the string method (Ref. [4] and Sect. 2.3.5).

M. Louwerse, *Efficient Control and Spontaneous Transitions*, Springer Theses,
https://doi.org/10.1007/978-3-031-40534-1

Calculating both time- and fully optimized protocols requires knowledge of the 4×4 friction matrix in relevant regions of control-parameter space. Each component of the friction matrix and its gradient (which is required for fully optimized protocols) with respect to each control parameter are interpolated using four-dimensional cubic splines [5, 6], which solve piecewise cubic polynomials over the data array to ensure smooth first and second derivatives. Since components of the friction matrix vary several orders of magnitude throughout control-parameter space, cubic splines are fit to the natural logarithm of each matrix component and re-exponentiated interpolated values from this surface to recover the friction coefficient. The natural logarithm of the friction is defined for this Ising model since each friction matrix component is positive. In low-dissipation regions at relatively large field values (top and left edges of the 2D control-parameter space in Fig. 4.2), noise causes the friction estimate at some control-parameter values to become negative; these were set to a constant value of $10^{-5}\,\tau_{\mathrm{rel}}/k_{\mathrm{B}}T$.

B.2.1 Time-Optimized Protocols

To calculate the time-optimized protocols (Fig. 4.3, middle), the friction is interpolated at 121 evenly spaced points along the fully naive protocol, then these points are reparameterized to be evenly spaced in terms of the 1D friction $\zeta^{\mathrm{1D}}(\boldsymbol{\lambda}) \equiv \sum_{i,j} \zeta_{ij}(\boldsymbol{\lambda})$. A reparameterization scheme commonly used with the string method was used to keep string points evenly spaced [7], with the modification that distances between discrete points are measured using the friction metric [8],to ensure constant excess power along this protocol (Fig. 4.11).

The string method divides a protocol of duration Δt into M discrete control-parameter points, with fixed time $\delta t = \Delta t/(M-1)$ between each string point and boundary conditions $\lambda(0) = \lambda_{\mathrm{initial}}$ and $\lambda(\Delta t) = \lambda_{\mathrm{final}}$. Since the minimum-work protocols are independent of duration (simply differing by a rescaling of δt), $\Delta t = 1$ without loss of generality. Let $\mathcal{L}^2(m)$ denote the length of the string up to discrete point $m \in [0, M-1]$. The generalized friction is used to define the distance between adjacent string points

$$
\begin{aligned}
\mathrm{d}\ell^2_{\boldsymbol{\lambda}(m),\boldsymbol{\lambda}(m-1)} = \sum_{ij} \left[\boldsymbol{\lambda}_i(m) - \boldsymbol{\lambda}_i(m-1)\right] \frac{\left[\zeta_{ij}\big(\boldsymbol{\lambda}(m)\big) + \zeta_{ij}\big(\boldsymbol{\lambda}(m-1)\big)\right]}{2} \\
\times \left[\boldsymbol{\lambda}_j(m) - \boldsymbol{\lambda}_j(m-1)\right] ,
\end{aligned}
\tag{B.1}
$$

so that the length of the string up to point m is

$$
\mathcal{L}^2(m) = \sum_{m'=1}^{m} \mathrm{d}\ell^2_{\boldsymbol{\lambda}(m'),\boldsymbol{\lambda}(m'-1)} , \tag{B.2}
$$

and estimates (up to a factor of protocol duration) the excess power on the system up to time $m\delta t$ in the protocol, with $\mathcal{L}^2(M-1)$ estimating the squared metric distance (and excess power) between protocol endpoints.

The string points $\boldsymbol{\lambda}^*(m)$ are then updated by linear interpolation of the current string so that the mth point is situated in proportion to the total distance $m/(M-1)$. Define $s(m) = [m/(M-1)]\mathcal{L}^2(M-1)$ as the distance along the string where point m should sit, then find the string point k in the original string such that $\mathcal{L}^2(k-1) < s(m) \leq \mathcal{L}^2(k)$. The reparameterized point $\boldsymbol{\lambda}^*(m)$ is then linearly interpolated between $\boldsymbol{\lambda}(k-1)$ and $\boldsymbol{\lambda}(k)$ by

$$\boldsymbol{\lambda}^*(m) = \boldsymbol{\lambda}(k-1) + \frac{s(m) - \mathcal{L}^2(k-1)}{\mathrm{d}\ell^2_{\boldsymbol{\lambda}(k),\boldsymbol{\lambda}(k-1)}}\left[\boldsymbol{\lambda}(k) - \boldsymbol{\lambda}(k-1)\right] . \tag{B.3}$$

The new string has points that are evenly spaced in terms of square distance (B.1), and therefore has constant excess power.

B.2.2 Fully Optimized Protocols

To calculate the fully optimized protocols (Fig. 4.3, bottom), the multidimensional Euler-Lagrange equation [9] for excess work (2.64) is solved numerically using the string method, as developed in Ref. [4]. This involves numerically solving the Euler-Lagrange equation for the protocol minimizing the excess work, by dividing the protocol into a discrete set of points along a "string" that runs between fixed protocol endpoints. The string is initialized as the naive protocol, where each point is evenly separated in Euclidean space. The string is updated from the nth to $(n+1)$th iteration by solving the set of linear equations for each component λ_α at each discrete point m along the string:

$$\begin{aligned} h_\alpha^{n+1}(m) - h_\alpha^n(m) = \Delta r \Bigg(&\mathrm{D}^2\lambda_\alpha^{n+1}(m) \\ &+ \sum_{ijk} [\zeta^{-1}]^n_{\alpha k}(m)\, \mathrm{D}h_i^n(m)\, \mathrm{D}h_j^n(m) \left[\partial_i \zeta_{kj}(m) - \frac{1}{2}\partial_k \zeta_{ij}(m)\right]\Bigg) , \end{aligned} \tag{B.4}$$

where

$$\mathrm{D}\lambda_\alpha(m) \equiv \frac{\lambda_\alpha(m+1) - \lambda_\alpha(m-1)}{2\delta t} \tag{B.5a}$$

$$\mathrm{D}^2\lambda_\alpha(m) \equiv \frac{\lambda_\alpha(m+1) + \lambda_\alpha(m-1) - 2\lambda_\alpha(m)}{\delta t^2} \tag{B.5b}$$

are finite-difference estimators of time derivatives for the string and δt is the difference in scaled protocol time between adjacent string points. Δr is a parameter controlling the size of the string update; $\Delta r = 10^{-4}$ for the 4D protocol and $\Delta r = 10^{-5}$ for the 2D protocol allowed convergence of each string to a constant-excess-power protocol (required for the time-optimized and fully optimized protocols). Unlike Ref. [4], the string was not reparameterized to keep points equally spaced; omitting this step allows simultaneously optimization of the spatial and temporal aspects of the fully optimized protocols.

The algorithm requires evaluation of the friction matrix and its spatial gradients with respect to control parameters at arbitrary control-parameter values, which are obtained from the spline fit. Ref. [4] assumes the integral relaxation time is constant and therefore the friction matrix can be replaced by the force covariance matrix; its derivative is simply the third cumulant of conjugate forces, $\langle \delta X_i \delta X_j \delta X_k \rangle$. This assumption greatly simplifies the calculation, but is not a good approximation for this system, as seen in Fig. 4.2 where the variation in relaxation time throughout the 2D control-parameter space is on the same order of magnitude as the variation in force covariance, and therefore variation of each element contributes significantly to the variation of the total friction. Optimization was carried out using 121 discrete string points while the symmetry of the protocol was constrained, $(\lambda_{\text{red}}(m), \lambda_{\text{blue}}(m), \lambda_{\text{green}}(m), \lambda_{\text{black}}(m)) = (-\lambda_{\text{red}}(M-1-m), -\lambda_{\text{green}}(M-1-m), -\lambda_{\text{blue}}(M-1-m), -\lambda_{\text{black}}(M-1-m))$. The 4D fully optimized protocol was passed through a Gaussian filter [3] with $\sigma = 0.03\Delta t$ to smooth noisy fields near protocol endpoints resulting from noisy friction estimates in low-dissipation regions. The smoothing affects the corresponding linear-response approximation to excess power in Fig. 4.11, which visibly dips for green and blue fields close to the start and end of the protocol, respectively. However, the total excess power throughout the vast majority of the protocol is constant as expected.

B.3 Protocol Simulation and Analysis

To simulate the control protocol, a trajectory is initialized from the equilibrium distribution at $\boldsymbol{\lambda}(t = 0)$, then its spin-flip dynamics are calculated for duration Δt using Glauber transition rates 2.6, while changing the control parameters according to the protocol. $N = 5000$ repetitions of each driving protocol were used to compute nonequilibrium averages of power, work, spin magnetization, total energy, entropy, and free energy. Quasistatic properties are calculated from Boltzmann-weighted equilibrium distributions along the protocol.

References

1. Louwerse, M.D.: Thesis code for 2d ising model (2022). https://github.com/mdlouwerse/MDL_thesis_code
2. Glauber, R.J.: Time-dependent statistics of the Ising model. J. Math. Phys. **4**(2), 294–307 (1963)
3. Press, W., Teukolksy, S.A., Vetterline, W.T., Flannery, B.P.: Numerical Recipes: The Art of Scientific Computing. Cambridge University Press, Cambridge (2007)
4. Rotskoff, G.M., Crooks, G.E., Vanden-Eijnden, E.: A geometric approach to optimal nonequilibrium control: minimizing dissipation in nanomagnetic spin systems. Phys. Rev. E **95**, 012148 (2017)
5. Walker, P.A.: Quadcubic interpolation: a four-dimensional spline method (2019). arXiv:1904.09869.
6. Walker, P.A., Krohn, U., Carty, D.: ARBTools: a tricubic spline interpolator for three-dimensional scalar or vector fields. J. Open Res. Softw. **7**(1), 12 (2019)
7. Maragliano, L., Fischer, A., Vanden-Eijnden, E., Ciccotti, G.: String method in collective variables: minimum free energy paths and isocommittor surfaces. J. Chem. Phys. **125**(2), 024106 (2006)
8. Rotskoff, G.M., Crooks, G.E.: Optimal control in nonequilibrium systems: dynamic Riemannian geometry of the Ising model. Phys. Rev. E **92**, 060102 (2015)
9. Morin, D.: Introduction to Classical Mechanics. Cambridge University Press, Cambridge (2008)